SHARING the SKIES

SHARING THE SKIES
NAVAJO ASTRONOMY

NANCY C. MARYBOY, PhD
AND DAVID BEGAY, PhD

Original concept co-designed by Indigenous Education Institute, Bluff, Utah,
and World Hope Foundation, Boulder, Colorado

RIO NUEVO PUBLISHERS
TUCSON, ARIZONA

CONTENTS

FOREWORD

For generations, the Diné People have observed the night sky, from which they developed a sophisticated philosophy and complex astronomy. This ancient knowledge is recaptured in this book in a friendly manner that is accessible to students and families.

The value of the comparative astronomy approach cannot be overestimated. Navajos should not only know their own astronomy, they should also learn the stories behind the Greek constellations, used in the schools today, as well as the space science that has guided space exploration and the collection of remarkable images from telescopes such as the Hubble Telescope.

The Navajo star paintings illuminate cultural knowledge that can be utilized in classrooms as well as by families, to provide relevance and comparative connections to Western classrooms, curriculum, and pedagogy.

The stories and information contained in this book will not only revitalize Navajo astronomical knowledge but can also be retold and applied in a contemporary way in schools using the NASA images and space science concepts. The juxtaposition of science, social science, and Diné cultural standards is innovative and most useful.

This is a most valuable resource for teachers and students and, in fact, anyone who has an interest in Yadiłhił (the Night Sky). The interconnection of Navajo astronomy, Greek mythology, and space science in one book has not been attempted before on this scale. I recommend that this book have a place in every school and home on the Navajo Nation.

The Hubble photos certainly provide humans the tool of imagination and cosmic feeling that heretofore may have been hard to achieve. One can only image how interconnected we are with the Universe from the photos, but it is so hard to put into words...because it is beyond words.

—NAVAJO ELDER

LELAND LEONARD
EXECUTIVE DIRECTOR,
DEPARTMENT OF DINÉ EDUCATION
NAVAJO NATION
WINDOW ROCK, ARIZONA

HOW TO USE THIS BOOK

FAR RIGHT: This graphic design depicts the worldview of the Navajo. There are four worlds shown, from which Navajos believe their ancestors emerged. These are colored black, blue, yellow, and white. There are four sacred mountains shown, which delineate the traditional Navajo country. There is a traditional Navajo hogan, or home, in the center, symbolizing the first Navajo home in the modern world. Four traditional plants are depicted within the four sacred mountains: corn, beans, squash, and tobacco. Four spiritual beings surround the perimeters of Navajoland. Navajos believe that light preceded all processes of life, and the light is illustrated as starlight coming through the four worlds. In the sky above, many significant Navajo constellations are illustrated. Conceptualized by David Begay and Nancy C. Maryboy. Design by Ken Grett.

This book is written primarily for Diné (Navajo) students, teachers, and families, so they can begin to understand their rich and complex heritage. It is also written for people of other indigenous heritages, so they can discover similarities between the Navajo astronomy and their own ways of seeing and knowing the skies. And finally, this book is written for the general public, of all ages and heritages, to share the Navajo ways of knowing the sky. We hope that this book will help to promote the sense of wonder and awe that we all feel as we gaze at the sky overhead, day and night, summer and winter, following the age-old patterns of the stars and other cosmic energies.

This book is not a textbook. It has been designed as a resource for teachers in the classroom as well as a book that families can read together. The book can be used for grades four through nine, and addresses middle school science and astronomy education standards. This book is also appropriate for high school and post-secondary classes featuring indigenous astronomy. The book fills a gap because nothing like it exists anywhere. In the book, stories are told through text, paintings, and photographs. The Navajo paintings were created by Melvin Bainbridge, a Navajo artist who lives on the Navajo Nation. We worked closely with him to ensure that the paintings would express the traditional Navajo teachings and that every detail would accurately depict Navajo constellations as seen by knowledge holders.

Our knowledge of the Greek astronomy comes primarily from historical writings, and, we have included photographs from the NASA Hubble, SOHO, GALEX, and ISS Missions, and from the Gemini Observatory. These amazing photos show stars and galaxies one could never see with the unaided eye, showcasing the scientific advances of our times, expanding the boundaries of what is known of the Universe.

We hope this book will be a useful resource for classroom and families. Stargazing can unite all generations of a family, from grandparents to the smallest child.

NAVAJO UNIVERSE

Characteristics of Yadiłhił (the Navajo Sky)

Ma'ii Bizǫ' (Coyote Star) is in the South, on the right side of the poster, as traditionally observed from Diné Bikéyah (Navajo Country)

Haashch'éshzhini (Black Spirit) and Coyote play important roles in the creation of the constellations

The Sun and the Moon are important parts of the Diné Universe

In the star map below, the primary Diné (Navajo) constellations are illustrated. Find the number on the star map to see approximately where each constellation is in the sky.

1. Náhookǫs Bi'ka' - Male Revolving One (includes parts of Big Dipper and Ursa Major)

2. Náhookǫs Bi'áád - Female Revolving One (includes Cassiopeia)

3. Náhookǫs Bikǫ' - Central Fire of Náhookǫs (North Star/Polaris)

4. Dilyéhé - Seed-like Sparkles (Pleiades)

5. Átsé Ets'ózí - First Slender One (includes Orion)

6. Hastiin Sik'aí'ii - Man with Solid Stance (includes Corvus)

17. Tł'éhonaa'éí (Moon)

16. Jo'hanaa'éí (Sun)

15. Haashch'éshzhiní (Black Spirit) - Representative of the whole night sky with Dilyéhé on the right temple and the Crescent and Full Moon on the mask's face

14a. Ma'ii (Coyote) tosses the stars, providing chaos

14. Ma'ii Bizǫ' - Coyote Star (Canopus)

13. Shash - Bear (approximately Sagittarius)

12. Ii'ni - Thunderbird (includes Pegasus for Ii'ni's body and six stars for its feather, beginning with Denebola in Leo)

7. Átsé Etsoh - First Big One (upper part of Scorpius)

8. Gah Hahat'ee - Rabbit Tracks (lower curved hook of Scorpius)

9. Yikáísdáhá - Awaits the Dawn (Milky Way)

10. Tsetah Dibé - Mountain Sheep (includes Beehive Cluster in Cancer)

11. Tłish Tsoh - Big Snake (includes parts of Puppis and Canis Major)

© 2012 Indigenous Education Institute www.indigenouseeducation.org www.sharingtheskies.com

PART I

INTRODUCTION

INTRODUCTION TO NAVAJO ASTRONOMY

AN OVERVIEW OF NAVAJO COSMOLOGY

Navajos have been living in the Four Corners region of the American Southwest for hundreds of years. The traditional land of the Navajo, Diné Bikéyah, includes parts of Arizona, Colorado, New Mexico, and Utah. The Navajo Reservation is approximately the size of West Virginia, or about 27,000 square miles. The population of the Navajo Nation is well over a quarter of a million people. Navajos have always believed that their homeland is geographically and spiritually located within an area delineated by four sacred mountains located in Arizona, Colorado, and New Mexico. Today, Navajoland, held in trust by the United States government, has been set aside by Treaty and Executive Order as an Indian Reservation. The Navajo, as a sovereign nation, have a treaty with the U.S. government.

HISTORICAL AND COSMOLOGICAL ORIGINS

Traditionally and historically, Navajos refer to themselves as the Diné, meaning, "The People." Sacred stories passed down from generation to generation tell of earthly and cosmological origins and relationships with continuous historical evolution through four worlds, ultimately leading to an Emergence that brought the Navajos to their present location.

Along with the traditional knowledge of evolution through the four worlds, higher Navajo consciousness acknowledges the origin of life through light, which preceded and gave birth to the evolutionary process. Thus it was a combination of biological and metaphysical processes that manifested as life on Earth.

The ancient origin of light provided the seed of consciousness and knowledge that is still acknowledged today in traditional Navajo society.

The origins of the Navajo go far back in time. Stories tell of the beginnings of life coming through interactions of light and darkness. Over thousands of years, insect-like beings inhabited the First World. Later, birds, animals, and finally, human beings inhabited the Second, Third, and Fourth Worlds. They eventually emerged into this present world, the Fifth World, where they live today in their homeland in the Southwest.

Map of Diné Bikéyah, Navajoland.

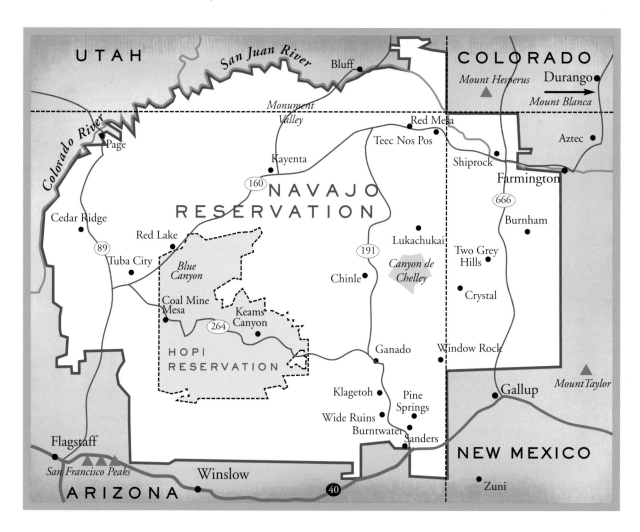

Navajos are related to other Athabaskan-speaking peoples from Alaska to Mexico. The language group includes Apaches, Northern Athabaskans, and Pacific Coast Athabaskan peoples. There is a distant language relationship to the Haida and Tlingit peoples as well.

Navajo Astronomy

Teaching and learning about Navajo astronomy is a considerably different process than studying Western astronomy and space science. The very time and medium in which knowledge can be transmitted is different between Navajo ways of knowing and scientific knowledge transmission.

Unlike conventional academic learning, which comes primarily from books, Navajo star knowledge has been passed down for countless generations through oral communication. Native star knowledge comes primarily through oral stories, field work, and interviews, which can take years to collect and verify, due to the esoteric nature of the knowledge and the method of communication.

The Ant Nebula.

Pieces of the stories are held by different families and clans, with acknowledged variations in interpretation. There is no one correct way to tell the star stories, since each version holds its own integrity and is connected to a specific lineage, often related to Navajo healing ways.

Today, very few people, even among the Navajos themselves, have a deep knowledge of Navajo astronomy. Like many other indigenous nations, much of the traditional knowledge is held by a few, and even this is only a portion of what was once known.

Traditional stories of the Night Sky were often spoken aloud with the enhancement of vocal performance, movement, and animal sounds. Most teaching traditionally took place during the winter months of October to February among family and clan members. Teaching associated with the Night Sky were shared within the traditional hogan, which itself was modeled and constructed in alignment with cosmic directions and principles. Navajo cosmology reflects the emphasis that Navajos place on the Night Sky and its interconnection with the Earth.

According to traditional Navajo protocol, cultural stories of the Night Sky, including stories of the Sun, the Moon, and the stars, can be told only during the winter months. However, at midsummer, a shorter version of the stories

Traditional Navajo hogan.
Photo by Nancy C. Maryboy.
Image design by Troy D. Cline.

can be shared for educational or healing purposes, during the time of the Summer Solstice and immediately thereafter.

Navajo stories of the Night Sky, called Winter Stories by the Navajo, are told from October (Ghą́ą́jı́'), when winter traditionally begins, to late February (Atsábiyáázh). The coming of the first thunder signifies the emergence of spring, at which point the Winter Stories are no longer told. The time of Winter Stories is considered to be a time of sharing and reflection when bears, reptiles, and insects are hibernating, and while plants are regenerating their potency for their next life cycle in the spring.

Although this book will be available year-round, we encourage teachers to be sensitive to the cultural protocol and use this book primarily during winter months. This provides a respect for the Navajo people as well as ensuring proper knowledge transmission.

ORGANIZATION OF STAR KNOWLEDGE

Navajos have organized their star knowledge from Diné Bikéyah, a central location in the American Southwest defined by the four sacred mountains. The order of the Navajo constellations is related to the geographical information provided from the cardinal directions of the sacred mountains. The cardinal directions, in turn, are directly connected to cosmic stellar processes as observed from this position of centrality in Diné Bikéyah. For example, east is the place where the Sun rises and west is where the Sun sets. These directions are also related to the equinox and solstice cycles of the Sun.

Embedded in the Navajo language, all directions have a clear relationship with cosmic processes. Ha'a'aah, East, means "Where the Sun comes up." Shá-diaah, South, means, "As the Sun travels with and for me." Ee'ee'aah, West, means, "Where the Sun goes down." Náhookǫs, North, refers to the motion of the circumpolar Náhookǫs constellations as they rotate while traveling around the North Star, Polaris.

The North Star itself is called Náhookǫs Bikǫ, the Central Fire. Náhookǫs Bi'kạ', the Male Revolving One, and Náhookǫs Bi'áád, the Female Revolving One, are situated on either side of Náhookǫs Bikǫ, and their constant circular pattern around Polaris, as seen from Earth, is referred to as circumpolar motion.

According to Navajo tradition, the Náhookǫs constellations are thought of as one complete constellation, not three separate star groups.

COSMIC VISION AND SPIRITUAL WORLDVIEW

The study of Navajo astronomy is very complex. It is highly cosmic and reflects a holistic worldview and way of life. The Navajo world-view includes an ordered Universe where everything is interrelated and all the pieces of the Universe are enfolded within the whole. At the same time, every piece contains the entire Universe, somewhat like a hologram, thus creating a network of relationship and processes in constant motion.

In accordance with this worldview, traditional Navajo astronomy is highly spiritual. The entire Universe is considered to be a living organism, a sacred organism, existing in a non-static and constantly regenerating process. The human is an integral participant within the dynamic whole. Every human action is considered cosmic and affects the web of universal relationship.

Eight primary constellations in traditional order.

Birth
Náhookǫs Bi'kạ'
Náhookǫs Bi'áád

Childhood
Dilyhéhé
Átse Ets'ózí

Maturity
Hastiin Sik'aí'ií
Átsé Etsoh

Old Age
Gah Hahat' ee
Yikáísdáhá

What impacts one impacts all. Everything in the Universe is also considered to be living and sacred. The human being is only a small part of a much larger world. Everything is seen to be constantly moving, and most patterns of movement are cyclical or circular. Chaos is thought of as just part of the natural order.

CONSTELLATIONS PROVIDE GUIDANCE AND VALUES

Navajo relationships with the stars can be very personal. Star constellations can be utilized for healing the body, mind, and spirit. Many Navajo constellations are depicted in human form, providing principles and values for living. The Náhookǫs constellations represent family relationships, similar to that of a mother and father in their home, providing warmth, stability, and security.

Other Navajo constellations depicted in human form represent differing stages of life, such as birth, childhood, maturity, and old age, while the various stages are tied to traditional values and the development of wisdom. The diagram on page 15 shows some of the constellations related to the stages of life.

RELATIONSHIP WITH ANIMALS AND NATURE

Many Navajo constellations are directly related to animals. Constellations such as the Mountain Sheep, Big Snake, Bear, Coyote, and the Horse (carrier of the Sun and Moon) illustrate attributes of the animals and show how animals interact with human life. Coyote occupies a special place within Navajo cosmology as a trickster and one who provides balance. Although Coyote often creates chaos, at the same time, on a larger plane, he provides harmony and balance.

Other Navajo constellations include natural elements such as Sun, Moon, and Thunder in the form of a Thunderbird. Although these may seem distant from one another, at a more complex level of understanding, they are interrelated and do not exist apart from each other. They connect through vital energies and vibrations.

The stars are also closely related to the growth of plants as well as to animal life processes such as mating and birth. Certain constellations, such as the Pleiades, give guidance of when to plant and when to stop planting. The phases of the Moon are also important guides for planting seeds and are related to reproductive cycles.

Traditional Navajo basket representing the Diné Universe.

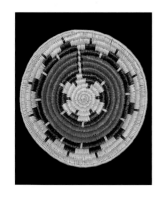

These seasonal growth processes follow a natural cosmic order as observed and experienced from Diné Bikéyah. The cycles of the stars, as well as the Sun and Moon, provide circular structure and order, which in turn are closely tied to the design and structure of the traditional Navajo round home, or hogan.

Map of the Universe.

Contemporary Navajo
basket: thirteen Yeis and Star.

ORDER PROVIDED BY COSMIC PROCESS

Eight main Navajo constellations, paired into twos, are traditionally ordered in sequence, beginning with the Náhookǫs constellation group and ending with the Milky Way, Yikáísdáhá. We will feature these main eight and several more constellations in this book, even though there are many more Navajo constellations in the night sky.

The constellation locations and names given are approximations. Most Navajo constellations are larger, though in some cases smaller, than the Greek equivalents. We have identified the Greek equivalents in order to show approximately where the Navajo constellations are located in the sky. In a few instances, we have not located the constellation with precision, in accordance with the wishes of our Navajo elders.

SPECIAL ATTRIBUTES OF STARS

It is interesting to note that in Navajo astronomy, each constellation is connected to an igniter star that provides light and identity to the group of stars. In addition, most Navajo constellations have a feather which signifies the spiritual essence of the constellation and the link to the spiritual wholeness of the Universe. A good example of the feather is found in the constellation of the Thunderbird, Ii'ni, in which the feather itself consists of six stars. The entire feather takes six months to become visible in the night sky.

This overview is but a brief glimpse of the skies through Navajo eyes. It is our hope that your journey into the night skies will take you beyond the stories and information contained here—that you will spend time observing the night skies from your own location, and discover the sense of awe and wonder that native people and people from all backgrounds have experienced for thousands of years.

NAVAJO TRANSLATIONS

Much of this book has been conceptualized through the Navajo language and subsequently translated into English. Translation is always a complex process, especially when dealing with two very different worldviews, one of process and interconnection and one of reductionism and compartmentalization.

The term Black God has been used in many books to describe what is more properly an essence, not a god or deity in the Western sense. Thus we have used the term Black Spirit as a translation of the Navajo term Haashch'éshzhiní throughout the book.

Due to subtle regional differences, mostly in pronunciation, there is no standardized way to write the Navajo language. In this book, we have chosen to write the Navajo language as spoken today in the central part of the Navajo Reservation. The diacritical marks used here, as well as the non-diacritical pronunciation guides, reflect the spelling as taught through the local tribal college curriculum.

INTRODUCTION TO GREEK ASTRONOMY

WHY IS GREEK ASTRONOMY IMPORTANT TODAY?

When we learn about astronomy today, we still study Greek stories of the Sun, Moon, and stars, thousands of years after the Greeks flourished as an early Western civilization. Greek astronomy is important to Native Americans because so many of the constellations and names which are used today are based on the names given to groups of stars by the Greeks, which are connected to cultural stories of the Greek sky. We communicate our Native star stories by using the Greek names of constellations such as Orion and Scorpius to identify

The Erechtheion of Ancient Greece.

and locate Navajo constellations and stars. The English language, with its incorporation of Greek names, is used today as the common language of the skies.

Western astronomy also honors other civilizations that came before the Greeks and after the Greeks, with star names and stories, even though sometimes we unknowingly designate the star knowledge of various early Western civilizations as Greek mythology or Greek astronomy. This includes the great star knowledge of the ancient Egyptians, Sumerians, Babylonians, Romans, Arabs, Hindus, and Persians.

BEFORE THE GREEKS

Before the Greeks, astronomers from early Western civilizations conducted precise observations of the stars, planets, Sun, and Moon. In the lands of what we now call Iraq, for example, Sumerian and Babylonian astronomers gave us stories and observation-based astronomical knowledge that we still use today. Stories of Orion and Scorpius, for example, predate the Greek stories by hundreds of years. The Greeks incorporated Sumerian and Babylonian star knowledge into their own mythology and astronomy, and much of it has passed down to us today.

The Egyptian kingdoms existed thousands of years before the Greek civilization. Ancient Egyptians were excellent astronomers and used star knowledge to develop time-keeping systems and to create sacred architecture. They developed a yearly cycle of 360 days, divided by tens, into 36 decans or intervals, based on the cycles of the Sun. Their calendar included a 24-hour day, which we still use today. This precise astronomical knowledge was reflected in the architecture of the great pyramids. Certain openings in some of the pyramids seem to have been connected with the patterns of constellations, including Orion. Many of the myths and histories of ancient Egypt are reflected and connected to star constellations, and great importance was attached to the cycles of the Sun.

Many ancient peoples in China, India, North and South America, Australia, and the islands of the Pacific, including Hawaii and New Zealand, were also keen observers of the night skies. Today we can access some of this knowledge through our universities and libraries, as well as through the oral traditions that have been passed down through the millennia. Even ancient Western astronomy originated from oral tradition and was later written in Egyptian hieroglyphics, on Sumerian and Babylonian clay tablets, and in Arabic and in Indo-European languages, predecessors of modern English.

GREEK CIVILIZATION

The Greek knowledge of the cosmos developed out of a large civilization that surrounded the Aegean Sea, part of an even larger civilization that surrounded the Mediterranean Sea. This civilization flourished between 3000 B.C. and 1000 B.C. The Greeks further developed this knowledge and began to write it down in their own language, around 1200 B.C.

Many of the star names that we use today are based on the Greek language. For example, the various stars in a given constellation can be identified, in order of brightness, through names from the Greek alphabet. They follow the order of the letters in the Greek alphabet, which begins with the alpha and beta, similar to A and B in English. Therefore, Alpha Pegasi is the brightest star in the constellation Pegasus while Beta Pegasi is the second brightest star in the same constellation, and so forth.

Greek astronomy was based on the sky as seen from the land of the Mediterranean Sea, located in Southern Europe, just a few degrees north of the latitude of the Navajo Nation. The latitude of Athens, Greece, is at 38 degrees, while the center of the Navajo Nation is at about 36 degrees.

The Greeks were a seafaring people, surrounded by many islands. The importance of water and navigation can be seen in their mythology and stories. Many of their stars and constellations are connected to the seas and water, such as Canopus, a southern star which is named for a famous ship navigator.

The Greeks, much like indigenous peoples today, saw the cosmos as an ordered whole. Their concepts of time and space were founded on Earth-based observation of the planets, stars, Sun, and Moon. Their far-reaching empire was bound together by a shared view of the Universe.

Their origin stories were created around histories that were considered sacred. They used their astronomical knowledge for living, planting, and sailing, and as guides for morals and values by which to live. Many Greek star stories contain warnings of what could happen if one disregarded the laws of nature.

To the Greeks, violence was a common part of life and as such violence is often reflected in their star stories. There is also emphasis on mating, since to the Greeks, mating brought regeneration and new life into the world. Greek cosmology includes stories of near and far cultures, including the well-known fall of Troy, and battles with other civilizations scattered around the Mediterranean Sea.

AFTER THE GREEKS

Much of the Greek knowledge of the cosmos was passed down to the next great civilization that ruled much of the Mediterranean, the Roman Empire. The Romans emerged from Latin-speaking tribes who lived in what we now call Italy, in Southern Europe. The Romans adopted many of the Greek stories about gods and heroes and their exploits in the sky. They too told of long-ago times when one could travel between the realms of earth and sky. The Romans told similar stories but substituted the Greek names with Latin names. So Zeus, the ruler of the Greek pantheon of heroes, became Jupiter, the ruler of the Roman skies. Cassiopeia, the wife and queen of Zeus, became the Roman queen Hera. Selene, the Moon, became Luna, and Helios, the Sun, became Sol or Sun. In many cases the English language uses derivatives of the Latin names even today. We say Sunday, meaning the Sun's day. We talk about a lunar landing, meaning a spaceship landing on the Moon, using the term "lunar," which is reflective of the Latin term for the Moon, Luna.

Another one of the many words that make up the English language is the month of January. January was named for Janus, a Roman god who represented the beginning and ending of things, and was closely related to the cycles of the Sun. Julius Caesar, a Roman ruler, was a skilled astronomer, and using his knowledge of the cosmos, he re-ordered the calendar, beginning with January. The month of July is named after Julius Caesar, while

European armillary sphere, 18th century, demonstrates observations of the skies from an Earth-centered perspective.

August is named after his successor, the Emperor Augustus. We use these terms almost every day, reflecting our connection with the ancient Romans.

Our Western star and time-keeping knowledge also includes parts of Scandinavian star stories. We get our name Thursday from Thor's Day, a Scandinavian name for a hero who represented the thunder in the sky. The name Friday honors the Scandinavian feminine deity Frigga, wife of Odin, the supreme god in Scandinavian mythology.

After the empires of the Greeks and Romans fell, much of the knowledge of the skies was carried on by scholars in the Christian Church and by noted Arab astronomers. Many of our star names today reflect Arab connections. The Arabs developed a complex system of astronomy that is used in Muslim countries today. Early Arab scholars preserved and added to the ancient knowledge of science, particularly in areas of medicine, mathematics, and astronomy. In fact, we owe them a debt of gratitude for preserving so much knowledge and retaining a link between ancient Greek knowledge and current Western astronomy.

Over the millennia, people of all agesaround the world looked up at the sky in wonder. They studied the day skies dominated by the Sun, and the night skies, full of stars, planets, meteors, and comets. They pondered their position in the Universe and developed concepts and stories that helped them explain their relationship with the astronomical world. These stories often expressed natural cycles and patterns, based on centuries of careful observation. They correlated the seasons of the year, the growth of plants, the availability of food, the cycles of hunting and fishing, with the celestial cycles of the sky. Through oral and written tradition, they passed on this knowledge, so vital to their survival.

In early civilized society, star stories reflected human psychology and communal values, and were often expressed through the elements and energies of earth and sky. Throughout time, people such as the Greeks and Navajos told stories of heroic quests and forces of complementarity, or seeming opposites that were dynamically connected to one another. Through these stories they created understanding and passed on values to their descendants. These same stories and values are used today in Navajo and other cultural astronomies and societies, and they can even be found woven into modern, scientific explanations of our incredible cosmos.

INTRODUCTION TO SPACE SCIENCE

BY ISABEL HAWKINS, PHD
RESEARCH ASTRONOMER AND DIRECTOR OF SCIENCE EDUCATION
SPACE SCIENCES LABORATORY, UC BERKELEY

A HISTORICAL ENDEAVOR

Humans across all cultures have venerated, observed, and studied the Sun, the Moon, the planets, and the stars beyond for thousands of years. Ancient peoples were keen observers of the skies, and astronomies were developed that featured complex interdependencies and holistic ways of knowing. Great civi-

From the Hubble Space Telescope: spiral galaxy.

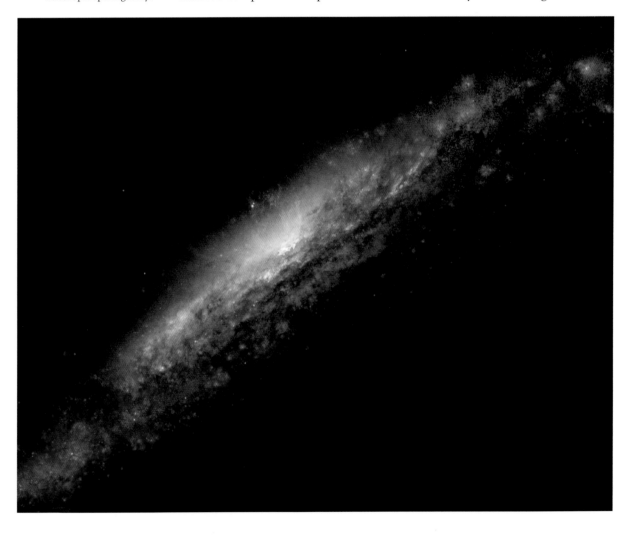

lizations, such as the Maya of Mesoamerica, developed intricate astronomical tracking systems and accurate calendars spanning many thousands of years. They built monumental structures aligned with cardinal directions to track the motions of celestial bodies in the sky for the purpose of planning and celebrating key dates in their ritual calendar, and for agriculture. Many of these astronomical traditions are still practiced today by the Maya of Mesoamerica.

Traditional farming communities time the cultivation of corn and other crops by observing the movements of the Sun and the stars. Similarly, indigenous peoples of the Southwestern United States, including Pueblo and Navajo, continue to incorporate knowledge of the stars in daily living, using keen observations of the motions of the Sun and stars for planting and ceremony. Astronomical knowledge and practices are reflected in ancient architecture such as the Great Houses found in Chaco Canyon, New Mexico, and many other ancestral Pueblo sites in other southwestern states including Arizona, Colorado, and Utah. Great cities were built over centuries that feature remarkable alignments with the cycles of the Sun and Moon.

Astronomy traditions continue to be practiced today by both indigenous and Western scientists. Scientists from NASA and other institutions conduct scientific research that emerges from questions about the characteristics, structure, and evolution of the Universe. Western research relies heavily on technologies and instrumentation that enhance the human ability of observation, combined with a mathematical theoretical framework. Satellites orbiting the Earth high above the atmosphere, like the Hubble Space Telescope, allow scientists to study the Universe in a different light. For example, instruments on the Hubble allow us to observe energetic ultraviolet light and x-rays emitted by distant objects such as planets, stars, and galaxies. Such energies of light are not able to penetrate the Earth's protecting atmosphere, since most of the light that comes through is what we call "visible light" or the colors of the rainbow. Satellite technology extends our ability to see astronomical objects in more detail than is afforded by visible energies, and we learn more about the characteristics of these objects.

However, the human eye of the professional astronomer rarely interacts directly with the light that reaches us from the stars across many thousands or millions of light years. The same technologies that extend our senses by allowing us to gather detailed information and data about the Universe also distance us from nature.

Professional astronomers observe stars, galaxies, and other distant objects in the sky using sophisticated instrumentation and computer imaging techniques mediated by telescopes made of several tons of steel. At times, as a professional astronomer, I have felt the need to leave the computer and walk outside of the observatory, to remind myself that I was inspired to become an astronomer because of the compelling beauty and mystery of the stars, the night sky, a beautiful full Moon, and the warmth of the Sun. Understanding our relationship as human beings with the objects of the Universe we are trying to study can help us be keener observers and more sensitive scientists.

Western science flourished during the European Renaissance, with discoveries of the laws that govern the motions of the planets and other bodies in the solar system, allowing us to calculate, for example, the orbits of the planets around the Sun. In the seventeenth century, scientists invented a new technology—the telescope. Galileo Galilei, the Italian scientist, was the first to point the telescope to the sky. Galileo made startling discoveries including spots on the Sun, the Moons of Jupiter, phases of the planet Venus, and craters on the Moon. He was also able to identify that the Milky Way was made of individual stars. His observations were crucial in supporting the theory that the Earth revolved around the Sun, and not the other way around.

Galileo was one of the first people in Europe who used instrumentation and experiments to study and understand the workings of nature. Perhaps more so than other accomplished scientists of that time, Galileo practiced science in the way Western scientists practice it today—through a constant interplay between mathematical theoretical models or descriptions, and testing of such models with experimentation and observation. The ability to predict natural phenomena through mathematically described theory, and to design experiments to test such theories, is the cornerstone of Western science.

RECENT SCIENTIFIC ADVANCES

Since the late 1950s, astronomy has evolved from observations done from the ground to space science observations done from space. This change came about due to our ability to launch satellites to study the Universe from above the absorbing quality of the Earth's atmosphere, and to send spacecraft to explore other planets in our solar system. More and more, space science is being done as a collaboration involving many countries, with satellites being built and launched into space by several countries around the world. Space science has uncovered that the Sun, planets, moons, asteroids, comets, and other objects in our solar system are mysterious and fascinating worlds different from Earth. Automated spacecraft and human-piloted expeditions to outer space and to the Moon have changed our understanding of the Universe.

Instruments that include cameras and other measuring devices aboard satellites and probes are making detailed observations of the Sun, uncovering a "wild" or stormy side of our nearest stars caused by magnetic forces. We know that the Sun provides us with heat and energy, and causes the seasons. The Sun is also responsible for space weather effects through a constant stream of charged particles that flow from the Sun called the solar wind. When we study the Sun with satellites, solar phenomena like flares and massive ejections of charged material into space become visible. The Sun causes auroras—Northern and Southern Lights—and affects astronaut safety, communication satellites, and power systems on Earth. In a society that increasingly relies on the use of technology, scientists need to understand space weather caused by the Sun.

International Space Station as seen from the space shuttle.

LIVING ON EARTH, EXPLORING SPACE

Our understanding of the Sun has grown tremendously over the past 100 years; however, it is only relatively recently that we recognized that knowledge of the Sun, and how it affects weather and long-term climate change such as global warming, is critical to protecting complex and fragile ecosystems on Earth. Earth science satellites with instruments pointing at the Earth are also studying the oceans, the landmasses, and the atmosphere from space, using techniques called "remote sensing." These instruments are contributing to our understanding of the ecology of our planet and the degradation that is occurring due to human activity and other factors.

While human exploration of space is still very risky, human presence in space is relatively common. Today, astronauts are living in space on a continual basis aboard the International Space Station. American astronauts have landed on the Moon. Spacecraft have also landed on Venus, Mars, the largest moon of Saturn, called Titan, an asteroid called Eros, and a comet called 9P/Tempel 1. Spacecraft have traveled to distant parts of the solar system, orbiting around or flying by all the planets except Pluto (now classified as a dwarf planet). Pieces of the distant worlds have been found in Antarctica, in the form of meteors from Mars or the asteroid belt. These meteors, as well as samples from the Moon brought back by the Apollo astronauts and samples of the solar wind brought back by the Genesis spacecraft, are being studied in laboratories.

Understanding other planets, moons, asteroids, and comets in the solar system can help us understand how the solar system formed 4.5 billion years ago, and how it evolved into what we see and investigate today. Such knowledge contributes to our understanding of how life may have evolved on Earth, and whether life is or was present in other parts of the solar system. Human activity in space requires a variety of skills and knowledge. Responsible exploration of space needs input from multiple perspectives and ways of knowing the Universe, including ethical considerations that take into account diverse worldviews.

WHAT WE'RE DOING TODAY

Exploration of the solar system from the NASA perspective is focused on the search for water, since water is the fundamental ingredient necessary for the

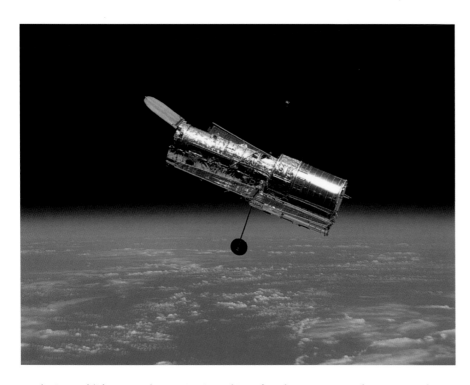

The Hubble Space Telescope.

evolution of life as we know it. Searching for the presence of water in planets like Mars can shed light into whether life has existed on Mars, and how life might have evolved on Earth. The discipline of "astro-biology" studies the conditions necessary for harboring life, including extreme environments such as hydrothermal vents at the bottom of the ocean, the planet Mars, and more remote areas of the solar system such as the moons of Jupiter. There is evidence, for example, that one of the moons of Jupiter, Europa, may harbor a liquid ocean deep beneath its surface, where life may exist.

Western astrophysics also focuses on studying the stars beyond our solar system. Over the past few years, astronomers have found hundreds of planets orbiting stars other than the Sun. These "exo-planets" may have Earth-like characteristics. Understanding how and where these exo-planets form and what they are like, and understanding how life survives under extreme conditions, are key steps in the search for life beyond Earth. Satellites are studying the most exotic and distant objects known, including black holes where gravity is so strong that not even light can escape, and quasars, the very brightest of galaxies, located many billions of light years away.

Violent and chaotic-looking mass of gas and dust is seen in this Hubble Space Telescope image of a nearby supernova remnant. Denoted N 63A, the object is the remains of a massive star that exploded, spewing its gaseous layers out into an already turbulent region.

Scientists are continuously searching for answers about the origin, nature, and evolution of the Universe. Most astronomers believe that the Universe had a fiery birth called the Big Bang 14 billion years ago, based on mathematical theories and observations. Space science is helping to answer what powered the Big Bang and what new forms of matter and energy the Universe is made of. Our atoms were created during the Big Bang and during successive generations of star birth and death. Each human being is as old as the Universe itself. Thus, a fundamental relationship between stars and human beings on Earth is the knowledge that the organic material in our bodies was created in stars. This knowledge is held by indigenous knowledge holders as well as Western astronomers.

In Western and indigenous astronomy, one of the most studied and revered star clusters in the sky is what Western scientists call the Pleiades and the Navajo call Dilyéhé. The Pleiades are venerated by indigenous peoples universally, who nurture a deep relationship with this sacred cluster. For example, the Cherokee origin stories talk about humans having come from the stars, specifically the Pleiades. On Earth, organic matter, what constitutes all living beings, is made from elements created in stars that exploded into the inter-stellar medium, providing the raw material for the Sun and solar system to form. Thus, the findings of Western science are consistent with the knowledge held by native astronomers, namely, that humans come from the stars. Humans are, indeed, stardust. Knowledge about how life originated on Earth is still a mystery to Western science, but we know that the chemical elements of life come from stars.

Other questions about the Earth and the Universe are still being studied, including regaining ecological balance on our planet, and the nature of different forms of matter and energy in the Universe. Many questions are still being asked, and many of the answers remain elusive. There is an opportunity for Western scientists and indigenous knowledge holders to work together to understand the relationship between the Earth, the Sun, and the Universe beyond.

Indigenous communities have always recognized interdependencies between the Sun and the Earth, and understanding the relationship between human beings and the Universe forms part of the enduring wisdom of indigenous people throughout the world. The strong and rich astronomical traditions that still thrive in indigenous communities can provide a framework for indigenous peoples to integrate Western astronomical knowledge with worldviews that are grounded in thousands of years of tradition. This book is an excellent example of the value of Navajo "ways of knowing," and I believe there is a need to re-awaken indigenous consciousness in all humans. What better way to do so than through astronomy and our yearning to understand and honor our place in the Universe!

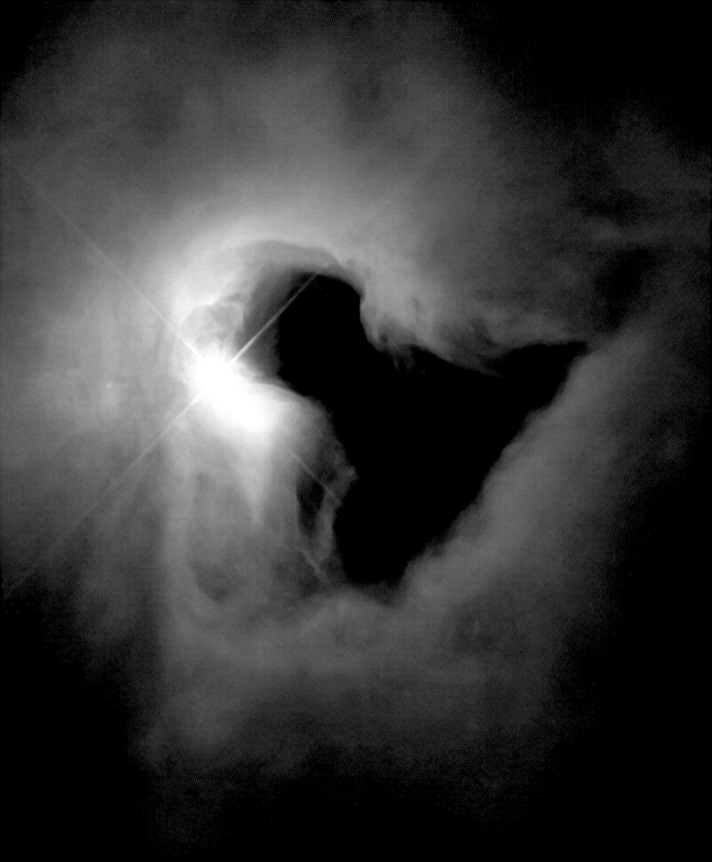

PART II

COMPARATIVE ASTRONOMY

LOOK UP FROM THE EARTH AT WHAT THE NAVAJOS CALL YADIŁHIŁ,
THE UPPER DARKNESS. . .

THE NÁHOOKǪS CONSTELLATION

NAVAJO NAME: Náhookǫs Constellation
PRONUNCIATION: Na hoe kos
TRANSLATION: Náhookǫs Constellation

WESTERN NAME: None
LOCATION: Big Dipper / Cassiopeia / Polaris
OBSERVED PATH: Circumpolar / Double Motion

The Náhookǫs Constellation is actually composed of three separate but related stars or groups of stars, which, to the Navajo, are thought of as one. Two of these, Náhookǫs Bi'ką' (Male Revolving One) and Náhookǫs Bi'áád (Female Revolving One), are located opposite one another, with the third, Náhookǫs Bikǫ' (Central Fire of Náhookǫs) in the middle. The orientation of the first two constellations changes, giving them the appearance of rotating slowly as they orbit (revolve) around Náhookǫs Bikǫ'. Because the three Náhookǫs star asterisms form one related group, they are often discussed together.

Náhookǫs Bi'ką' and Náhookǫs Bi'áád move in a double motion through the sky. First, both constellations follow a counter-clockwise orbit around Polaris. Second, they rotate around themselves. They also appear to move in relation to the horizon, at certain times appearing above Polaris, at others

The Big Dipper's motion around Polaris has been used as a calendar and an indicator of seasonal change. At roughly the same time throughout the year, you can see the Dipper start its nocturnal circuit in a different position relative to each season.

below. This movement can be used as a time guide, as the position of the constellations changes with the seasons as seen in the image on page 33.

NÁHOOKǪS BI'KĄ' / URSA MAJOR

NAVAJO NAME: Náhookǫs Bi'ką'
PRONUNCIATION: Na hoe kos Bih kah
TRANSLATION: Male Revolving One

WESTERN NAME: The Great Bear
LOCATION: Big Dipper / Ursa Major
OBSERVED PATH: Circumpolar / Double Motion

NÁHOOKǪS BI'KĄ'

The first constellation of the Náhookǫs group is Náhookǫs Bi'ką' (Male Revolving One). He is considered to be a male warrior, a leader and father, and sometimes a grandfather, who provides for his family. He protects his family with weapons such as the bow and arrow. He reflects the ideal characteristics of a provider and protector for his family, people, and home. From Diné Bikéyah

(Navajo Country), Náhookǫs Bi'ką' is seen in the sky every night, year-round, and thus provides direction, timekeeping, and stability for the other stars as well as for all living things.

Náhookǫs Bi'ką' is a circumpolar constellation, meaning that it orbits around Polaris. The word Náhookǫs also describes the circular turn (rotation) the constellation makes around itself. Additionally, the constellation's position changes with the seasons. On winter evenings, it can be seen on the right side of Polaris with the tail facing down. On summer evenings, it is on the left side with the tail facing up.

Many Native American tribes, including the Algonquians and Iroquois, as well as other European groups such as the Sami people of Norway, Sweden, and Russia, conceptualize the Big Dipper as an ongoing bear hunt. They tell of the Big Dipper as a female bear, who is roused out of her den at the end of her winter hibernation by a group of hunters, depicted by the stars in the handle of the dipper. The hunters pursue the bear all summer, finally catching up with her in the fall. The hunters shoot the bear and cook it in a pot.

In many of the stories, the spirit of the bear goes into a new female bear that is hibernating, and emerges as a baby female bear in the spring. The hunt then begins all over again.

Some tribes, including the Iroquois, identify the hunter stars as various birds, including the Robin, who is colored by the bear's blood and acquires a red breast. Other tribes say the drops of blood from the bear become the red leaves of fall.

URSA MAJOR

Ursa Major is known as the Great Bear and includes the Big Dipper. It is one of the most familiar and visible constellations from the Northern hemisphere. Even before the Romans named the star group Ursa Major, the Greeks called it Artos, meaning bear. From Artos comes the English word arctic, referring to the far North and cold areas.

In Greek myth, Callisto was a beautiful young girl who was discovered by Zeus while he was investigating the earthly destruction caused when the Sun briefly went out of control and came too close. Zeus put the Sun back into its correct path and checked the entire Earth for damage. When he came to the land of Arcadia, he discovered beautiful Callisto resting on the grass. Zeus fell in love with the young huntress, and together they had a son named Arcas.

Now Zeus had a very jealous wife, named Juno, and when Juno learned about Callisto and Arcas, she became enraged. She reached down from the sky, grabbed Callisto and dragged her around the ground. Callisto began to grow long shaggy hair, long pointed claws, and a pointed nose. She could no longer speak words and could only roar—she had become a bear.

Poor Callisto spent a year roaming around the forest, always keeping a protective eye on her son. One day, her son spied the bear sleeping and drew out his bow and arrow to shoot her, not knowing it was his mother. But Zeus, who still cared for Callisto, intervened, grabbing Callisto and bringing her up into the sky, where Callisto became a new constellation, the Great Bear. There she shines in all her glory, eternally infuriating Juno.

Countless other stories of the Big Dipper exist. Celtic people saw it as both a wagon and a bear. Arabs saw it as a funeral procession following a coffin. European farmers considered it to be a plow. Today most North Americans call it a dipper, due to its gourd-like shape.

NÁHOOKǪS BI'ÁÁD / CASSIOPEIA

NAVAJO NAME: Náhookǫs Bi'áád
PRONUNCIATION: Na hoe kos Bih aad
TRANSLATION: Female Revolving One

WESTERN NAME: Cassiopeia
LOCATION: Cassiopeia
OBSERVED PATH: Circumpolar / Double Motion

NÁHOOKǪS BI'ÁÁD

The second constellation within the greater Náhookǫs constellation is Náhookǫs Bi'áád, the Female Revolving One. She depicts the female companion of Náhookǫs Bi'ką'. She is considered to be a mother, and sometimes a grandmother, who exemplifies strength, motherhood, and regeneration.

She reflects the ideal characteristics of stability and peace in the home. She also provides for her family through her female weapons of a grinding stone and stirring stick, used to fight off hunger and ensure good nutrition and a healthy lifestyle for her family.

This constellation is also circumpolar, orbiting counter-clockwise around Polaris, while rotating around itself. Like Náhookǫs Bi'ką', it has different positions according to the seasons.

CASSIOPEIA

Cassiopeia is considered one of the circumpolar constellations as it revolves around the North Star, in a counterclockwise direction. It can be seen in the Milky Way and is always located opposite the Big Dipper, with the North Star between the two. In Greek stories, Cassiopeia is a beautiful queen who is punished by being tied to a chair for all eternity, doomed to rotate in her seat as she revolves around the North Star. Many ancient poets, including Homer, Herodotus, and Ovid, wrote about Cassiopeia.

The story of Queen Cassiopeia is very old, going back well beyond the Greeks, and containing aspects of legends from the Phoenicians, Assyrians, and Babylonians (from the lands of present-day Iraq). The ancient Phoenicians who sailed around the Mediterranean Sea called the queen Quassiopeaer, referring to her toned face. She was often thought to be a dark-skinned woman from the land of Ethiopia in Africa. Many years later, Arabs called the constellation the Hand of Fatima, meaning the daughter of Mohammed.

In even earlier times, Cassiopeia was the Great Earth Mother, Hecate, who was the ruler of the earth and sky, the epitome of creativity and life-giving forces. But as civilization changed, the positive aspects of her character were transformed into negative traits, and by the time of the Greeks, she was being punished for arrogance, willful pride, and vanity.

The story of Cassiopeia is part of a much larger, very ancient story centered around the warrior Perseus, the son of Zeus. Perseus wore a helmet designed to make him invisible to his enemies, and had winged sandals that could make him fly. He flew over a craggy beach one day and spotted a beautiful young girl chained to the rocks. She was Andromeda, the daughter of King Cepheus and Queen Cassiopeia. She had been chained to the rocks as a sacrifice to the Whale Monster Cetus as a punishment for her prideful mother, Cassiopeia.

Perseus immediately fell in love with the young Andromeda. He learned that he would have to kill the monster Cetus in order to free the girl. Using the head of another monster that he had already killed, Medusa, he tricked and killed the Whale Monster. When he tried to take Andromeda as his bride, her mother, Cassiopeia, betrayed him, and he had to use the head of Medusa to turn the queen to stone.

Later, Poseidon, the ruler of the oceans who supported Cassiopeia, placed the queen in the sky as a constellation. But because of her bad deeds, she was tied into a chair and made to revolve around the North Star every night. This meant that she would be upside-down for half of every revolution—not a pleasant way to spend eternity. Perseus did eventually marry Andromeda and later became the founder of the Persian kings who ruled the land we now call Iran.

NÁHOOKǪS BIKǪ' / POLARIS

NAVAJO NAME: Náhookǫs Bikǫ'
PRONUNCIATION: Na hoe kos Bih kwo
TRANSLATION: Central Fire

WESTERN NAME: North Star
LOCATION: Polaris
OBSERVED PATH: Stationary

NÁHOOKǪS BIKǪ'

This star represents the cosmic center of the Navajo Night Sky. It also represents the central fire of the Navajo family hogan (home), as it is located between Náhookǫs Bi'kạ' and Náhookǫs Bi'ááд. It connects the two complementary yet distinct constellations, which are situated opposite one another. Náhookǫs Bikǫ' is the star that never moves and thus provides direction and stability to the cosmic process.

In Navajo culture, the central fire in a traditional hogan has multiple values and therefore is extremely important. It provides stability for the family, as well as a sense of security and well-being. Náhookǫs Bikǫ' provides warmth and tranquility as well as a focus, place, and means for ceremonial healing.

Long ago the central fire was on the ground in the hogan, providing light and heat for cooking. Today most families use a metal wood-burning stove, though it is still located in the center of the hogan.

Náhookǫs Bikǫ' can also be connected to Ma'ii Bizǫ' (Canopus), the Coyote Star, which is located in the South, to establish a north-south connection.

From Diné Bikéyah, Náhookǫs Bikǫ' is not located directly overhead, but about halfway (roughly 35 degrees) above the horizon. However, in Alaska, it would be much higher, and below the equator, you wouldn't be able to see it at all.

POLARIS

Polaris has been the North Star (the pole star that indicates the direction north to observers) for almost 3,000 years, but it was not always the North Star.

Star trails around Polaris, above Mauna Kea, Hawaii.

Seven thousand years ago, around 5000 B.C., the constellation Draco (the Dragon) was in the north position. Later, around 2700 B.C., Thuban was the North Star, during the time of the Egyptians. Over the millennia, as the stars continue to change positions, different stars will become the North Star. As the North Star has changed, so have the constellations. When Draco ceased to be the North Star, astronomers redrew the constellations and removed the wings from the dragon. The wings turned into the body of Ursa Minor, the Little Bear. The tip of the tail of Ursa Minor is now Polaris, the current North Star.

Europeans call Ursa Minor the Little Dipper. This is not the same constellation as Náhookǫs Biʼááád, the Female Revolving One in Navajo, which is sometimes called the Female Dipper. Rather, Náhookǫs Biʼááád is Cassiopeia.

Ursa Minor is named for the Greek hero Arcas, meaning "north" or "bear," as in the English term "arctic." Arcas was the son of Callisto, the maiden who was ravished by Zeus while she was out hunting. Juno, the jealous wife of Zeus, punished Callisto by beating her up and turning her into a bear (see the story in Ursa Major on page 35).

An interesting anecdote about Polaris comes from desert nomads in the Middle East, who say that watching this star intently can cure itching eyes. Also, some indigenous Scandinavian people call Polaris the "Nail of the North."

DILYÉHÉ / PLEIADES

NAVAJO NAME: Dilyéhé
PRONUNCIATION: Dil yeh heh
TRANSLATION: Seed-like Sparkles

WESTERN NAME: Seven Sisters
LOCATION: Pleiades
OBSERVED PATH: East to West

DILYÉHÉ

Dilyéhé is an agro-astronomical constellation, meaning that it is an indicator for the beginning and ending of the planting season. In early May, Dilyéhé becomes invisible in the evening on the western horizon. For the previous month, Dilyéhé has been traveling with the Sun all day, and at first dark, when one would just be able to see the stars, it can be seen going down with the Sun.

The Navajos use the cycles of Dilyéhé to determine when to plant corn. The time when they were not able to see Dilyéhé, is typically when they would begin to plant.

The teachings of Dilyéhé are very important because if one plants too early in the spring, the young corn can be destroyed by late frost. Conversely, if one plants too late in the summer, an early fall frost can destroy the corn before it ripens. For traditional farmers, using this constellation to plant at the right time was a matter of survival. Because of this, much respect and significance is attached to following and living with the natural cycles of seasonal weather and planting related to Dilyéhé.

There are many stories about this constellation among the Navajo. The name Dilyéhé refers to the seed-like sparkles that are evident in the movement of these stars at night. Some say that the white spots on a young fawn represent Dilyéhé.

One Navajo story tells of seven little warrior boys (Beesh Ashiké—Hard Flint Boys) who perform healing activities in Dilyéhé-like formations during the summer months. Another story tells about a family that is engaged in planting in a Dilyéhé-like formation. Still another story tells of six little warrior boys who are practicing their skills with the bow and arrow, accompanied by a woman with a buckskin slung over her back. They wander off and disappear over a hill. When they disappear (the point when you can't see Dilyéhé in the sky), it is time to plant. When they re-emerge on the eastern horizon in late June and early July, it is time to stop planting.

In addition to the Navajo, many indigenous groups use or have used Dilyéhé as a planting guide, including the Incas of Peru.

PLEIADES

Stories of the Seven Sisters or Seven Brothers occur in most cultures around the world. It is interesting that many cultures also make reference to a lost sister or brother, or lost Pleiade. Some stories connect the lost Pleiade with the Big Dipper and say the lost, or stolen, one ended up in the handle of the Big Dipper.

Hubble: The Pleiades.

The Pleiades are actually many more than six or seven stars. They are located near the constellation Taurus the Bull, and appear to be running from Orion the Hunter. They twinkle and, although they are not bright, they are easily recognized. The Pleiades have been used by both Western and indigenous people throughout the ages as part of a celestial calendar for both agricultural cycles and navigational directions.

The Greek names of the seven sisters, beginning with the eldest, are Taygete, Merope, Alcyone, Celaeno, Electra, Asterope, and Maia. Some stories say Jupiter was the father of the girls.

One Greek story identifies the seventh Pleiade as Electra, who was the mother of Dardanus, the founder of Troy. When the city of Troy was destroyed, she was devastated. Electra mourned her relatives and descendants. She unbound her hair as a sign of her grief and moved far to the north. Once in a while she returns with her hair streaming behind her, as a great comet.

Another Greek story says that the Seven Sisters were daughters of Pleione, a young girl whose own parents were Thetis and Oceanus, water beings. In this version, six of the sisters were married to gods but the seventh, Merope, married a mortal and was thus shunned by the others. Out of embarrassment, this seventh sister did not shine as brightly as she did not wish to be seen.

Yet another story identifies the Seven Sisters as a flock of doves. In this story the girls were traveling with their mother when they were attacked by the hunter Orion, who was a mean giant. The girls prayed to the ruler of all gods, Zeus, who took pity on them and turned them into seven doves. They flew into the air, escaping the giant. Orion pursued the girls for seven years, until both he and the doves wound up in the sky. Even today, one can say that Orion is chasing the Pleiades through the night skies.

BARNARD'S MEROPE NEBULA, IC 349

NAME: Barnard's Merope Nebula, IC 349
DESCRIPTION: Reflection Nebula in the Pleiades
POSITION (J2000): R.A. 03h 46m 21.30s Dec. +23° 56'28.0"
DISTANCE: 380 light years
CONSTELLATION: Taurus

This apparition is actually an interstellar cloud caught in the process of destruction by a strong radiation from a nearby hot star. This haunting photo on page 44, taken by the Hubble telescope, shows a cloud illuminated by light from the bright star Merope, located in the Pleiades star cluster. This cloud is called IC 349, otherwise known as Barnard's Merope Nebula.

The nebula in the photo on page 44 is 3,400 astronomical units, or 0.0537 light year wide. The nebula is 3,500 astronomical units (0.06 light year) away from the bright star Merope, located at the upper right, just outside the photo.

1. How did the cloud get its shape?
The cloud has been shaped by its closeness to Merope. The distance between the two objects is about 3,500 times the separation of the Earth from the Sun (about 0.06 light year). The cloud, which has been drifting through the Pleiades star cluster, is moving closer to Merope at a speed of about 6.8 miles per second (11 kilometers per second). Astronomers have proposed that the strong starlight shining on the dust in the cloud decelerates the dust particles. Physicists call this phenomenon "radiation pressure."

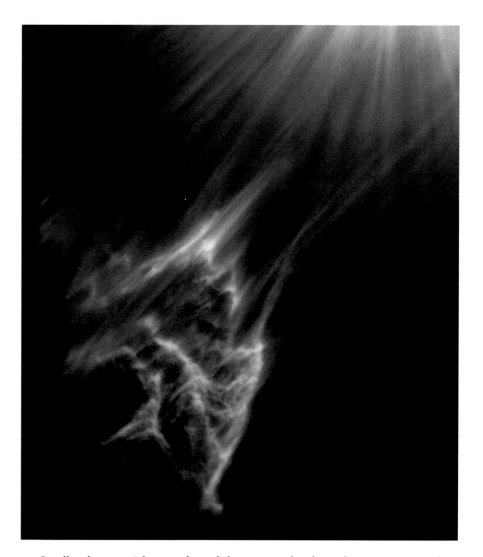

Hubble: Reflection of the
Nebula in Pleaides.

Smaller dust particles are slowed down more by the radiation pressure than the larger particles. Thus, as the cloud approaches the star, there is a sifting of particles by size, much like grain thrown in the air to separate wheat from chaff. The nearly straight lines pointing toward Merope are thus streams of larger particles, continuing on toward the star while the smaller decelerated particles are left behind at the lower left of the photo.

2. Where is the bright star Merope?
Merope is just outside the photo on the upper right. The colorful rays of light

at the bottom right, pointing back to the star, are an optical phenomenon produced within the telescope and are not real. However, the remarkable parallel wisps extending from lower left to upper right are real features, revealed for the first time through Hubble's sharp "eyes."

3. What is the cloud's fate?

Over the next few thousand years, the nebula—if it survives the close passage without being completely destroyed—will move on past Merope, somewhat like a comet swinging past our Sun. This chance collision allows astronomers to study interstellar material under very rare conditions and thus learn more about the structure of the dust lying between the stars.

4. Where can I find out more about this topic?

For more information from HubbleSite and beyond, visit the related links at: http://hubblesite.org/newscenter/newsdesk/archive/releases/2000/36/related/

ÁTSÉ ETS' ÓZÍ / ORION

NAVAJO NAME: Átsé Ets' ózí
PRONUNCIATION: A tseh ets osi
TRANSLATION: First Slender One

WESTERN NAME: Orion / The Hunter
LOCATION: Includes Orion
OBSERVED PATH: East to West

ÁTSÉ ETS' ÓZÍ

Átsé Ets' ózí is often envisioned as a young, strong warrior whose responsibility is to provide protection to his family and people. Implied in the name Átsé Ets' ózí is the cyclical emergence of the constellations in a vast, cosmic process.

Traditionally, this constellation is paired with Dilyéhé. The two constellations follow one another in a cyclical east-to-west pattern. Átsé Ets' ózí is also associated with Átsé Etsoh, First Big One (page

51). These two constellations are never seen at the same time in the night sky. When one is in the sky, the other is below the horizon and vice versa.

The relationship between the two constellations Átsé Ets' ózí and Átsé Etsoh is illustrated in a traditional story about in-laws. The story says that a mother-in-law and son-in-law should not see one another in daily life. In fact, a traditional Navajo mother-in-law might even wear a bell to warn the son-in-law of her approach. A similar relationship occurs between Orion and Scorpius.

ORION

The constellation of Orion is often considered a hunter. It is unique and easy to identify in the sky, with three bright stars making up his belt and another three hanging below his belt representing a sword. Cultures all over the world have stories about this constellation.

The ancient Greeks saw Orion as a kneeling warrior with his arm upraised, a hunter who was going to kill Taurus the Bull. In many drawings, his left foot is resting on Lepus the Hare. The bright star on his right shoulder is Betelgeuse.

Orion by Hevelius.

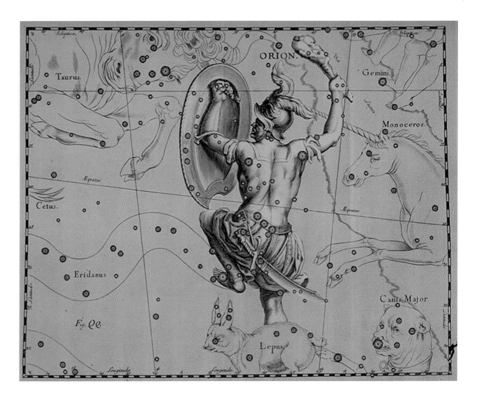

In ancient Arabic, Betelgeuse was seen a little lower than the shoulder and was called Ibt al-Jauzah, meaning "The Armpit of He Who Is in the Center."

The name Orion may have come from the Sumerian word Uru-ann, "Light of the Sky." Many cultures see this constellation as a warrior: strong, proud, vain, arrogant, and violent.

Orion was born in a rather strange way. He seems to have had no mother, except for the earth and the skin of an ox. Legend has it that Orion's father, Irieus, once welcomed three strangers into his home. The strangers happened to be the gods Jupiter, Neptune, and Mercury. They were very grateful to Irieus for his hospitality and said they would grant Irieus his dearest wish. Irieus said he most wished for a son. At that point, the three gods urinated onto an ox skin and told Irieus to bury it in the earth. Months later the baby Orion was born from the ox skin. Some say that his name was Urion (urine), later changed to Orion.

Orion grew up strong and large as a giant. He loved to kill things. One story tells of him assaulting Merope, the beautiful young daughter of King Oenopion. The father of Merope was enraged and had Orion blinded. Orion spent years trying to regain his sight. He traveled to the end of the Greek world, the island of Delos, and met Helios, the Sun. Helios fell in love with Orion and restored his sight. Helios and Orion spent the night together and when Eos, the Dawn, arrived in the morning, she was surprised to see them both, and blushed. That is why the sky is pink at dawn, according to the Greeks.

The goddess and huntress Diana also fell in love with Orion. Diana had a brother, Apollo, who was very close to Helios, the Sun. Apollo was jealous that Orion had spent the night on his island, Delos, and wanted to kill him. He decided to trick his sister. He called on the monster Scorpion to kill Orion. When Orion saw the Scorpion he was so frightened that he jumped into the sea to escape the monster.

Apollo knew Orion would flee into the sea and had invited his sister Diana to the shore with him. He teased her about her hunting skills and taunted her, saying he did not think she could hit the black dot out in the sea. Diana was angry and shot her arrow at the black dot, not knowing that she was killing the man she loved. Soon the body of Orion washed in with the waves and Diana was overcome with grief. Later, she put Orion and the Scorpion up in the sky, as bright constellations.

The tale of the Scorpion pursuing Orion is played out in the sky every night. When Orion sets in the West, the scorpion comes up in the Southeast. But the Scorpion never kills Orion. They never even meet, as the two constellations are never in the sky at the same time. This is very similar to the Navajo story of the mother-in-law and son-in-law, Átsé Ets' ózí and Átsé Etsoh, who are supposed to avoid each other.

Orion is also connected in ancient stories to Noah's Ark. The Sumerians, for example, have many tales of Orion as the son of Noah's niece. In these stories, he is called Gilgamesh and is a wild king who has many adventures.

HORSEHEAD NEBULA, BARNARD 33

NAME: Horsehead Nebula, Barnard 33
DESCRIPTION: Dark Nebula
POSITION (J2000): R.A. 05h 40m 59.00s Dec. -02° 37' 30.0"
DISTANCE: 490 pc (1,600 light years)
CONSTELLATION: Orion

Hubble: The Horsehead Nebula.

Rising from a sea of dust and gas like a giant seahorse, the Horsehead nebula is one of the most photographed objects in the sky.

The Horsehead, also known as Barnard 33, is a cold, dark cloud of gas and dust, silhouetted against the bright nebula, IC 434. The bright area at the top left edge is a young star still embedded in its nursery of gas and dust. Radiation from this hot star is eroding the stellar nursery. The top of the nebula is also being sculpted by radiation from a massive star located out of Hubble's field of view.

Only by chance does the nebula roughly resemble the head of a horse. Its unusual shape was first discovered on a photographic plate in the late 1800s. Located in the constellation Orion, the Horsehead is a cousin of the famous pillars of dust and gas known as the Eagle nebula. Both tower-like nebulas are cocoons of young stars.

The Horsehead nebula lies just south of the bright star Zeta Orionis, which is easily visible to the unaided eye as the left-hand star in the line of three that form Orion's Belt.

HASTIIN SIK' AÍ'IÍ / CORVUS

NAVAJO NAME: Hastiin Sik' aí'ií
PRONUNCIATION: Hasteen Sick eye ee
TRANSLATION: Man with a Solid Stance

WESTERN NAME: The Crow
LOCATION: Includes Corvus
OBSERVED PATH: East to West

HASTIIN SIK' AÍ'IÍ

This star formation depicts stability, strong posture, and balance in the form of a man standing with his feet apart in a solid stance. Hastiin Sik' aí'ií, consisting of more than thirty stars, is much larger than the Greek constellation Corvus, which occurs in approximately the same location. From Diné Bikéyah, Hastiin Sik' aí'ií looks like a man standing in the Southern part of the sky.

Hastiin Sik' aí'ií is generally associated with the division of the winter and summer seasons.

This constellation appears over the southeastern horizon in the early predawn during the latter part of September, when the weather turns colder and the first frost arrives; the color of the leaves on the trees starts to change and snow may appear at higher elevations.

At this time, the constellation is seen as a representation of the parting of the winter and summer seasons. Traditionally, the emergence of this constellation is an indicator that winter-related communal and social activities can begin.

CORVUS

The Greek constellation Corvus represents a crow or raven and is seen in the southern sky near the constellations Crater the wine container and Hydra the

Corvus by Hevelius.

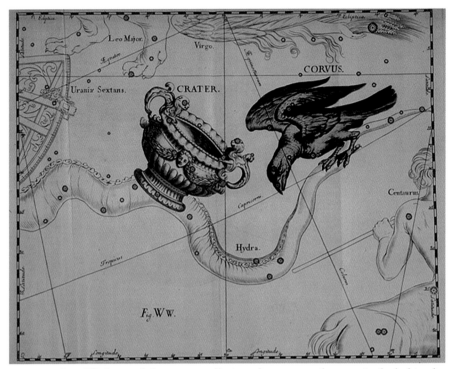

water snake. All three of these constellations have stars that are included in the Navajo constellation Hastiin Sik' ai'ií.

In the land of ancient Greece, there were certain women who could foretell the future by asking questions. They were called Sybils. The Sybils predicted the future by interpreting the song of certain birds. Among these birds were the crows, or ravens. Coronis, the goddess of the Sybils, eventually became the Greek goddess Athene. Apollo, the god of the Sun, married Athene and inherited the raven as his symbol along with the oracle of the Sybils. In those days the raven was a completely white bird, without a trace of black.

One day, Apollo was doing a ceremony to make a sacrifice to the ruler of all the Gods, Zeus. He needed some water from a special spring and sent his raven with a container to get the water. The container was called a crater, a special two-handled container in which Greeks would often mix wine and water so the wine would not be too strong. The raven flew off towards the special spring, but on the way he saw a tree full of figs that were almost, but not quite, ripened. The greedy raven really wanted the figs and decided to wait for them to ripen. After a few days they ripened and the raven ate to his content.

Full, and now feeling guilty, the raven knew that Apollo was going to be furious with him for being late with the water. So he grabbed a water snake (Hydra) in his mouth and flew back to Apollo, saying that the delay was the fault of the water snake. However, Apollo could see through the raven's ruse. Apollo became very angry, not only because he had to substitute other water for his sacrifice, but also because the raven tried to fool him. As punishment, Apollo changed the raven's beautiful white feathers into a sooty black color and placed him in the sky forever.

The black raven remains in the sky today. He is near the water jug but he cannot drink from it because the water snake is in the way. He must remain forever thirsty and his plight reminds people not to lie or shirk their responsibilities.

ÁTSÉ ETSOH

NAVAJO NAME: Átsé Etsoh
PRONUNCIATION: A Tseh etsoh
TRANSLATION: First Big One

WESTERN NAME: The Scorpion
LOCATION: Approximately upper Scorpius
OBSERVED PATH: East to West

ÁTSÉ ETSOH

Átsé Etsoh is located near the southern horizon of Diné Bikéyah. This constellation represents a Navajo elder standing with a cane, holding a basket of seeds. He exemplifies the wisdom that comes with age and the process of regeneration. His cane signifies strength and stability. The basket is representative of the entire cosmos while the seeds are the regenerative growth process that provides life.

Átsé Etsoh is a large constellation that includes the upper part of Scorpius as well as various surrounding stars. Traditional Navajos make reference to this constellation by its position in the night sky. In the early summer months of June and

July, Átsé Etsoh will appear as an upright "J." As summer turns into fall, the constellation will appear to tilt more to its side, until it is almost lying down, in alignment with the southern horizon. Navajos know that this is when the winter is near and hunting for deer can take place.

Implied in the term Átsé Etsoh is the concept of a cyclical cosmic emerging and evolving process. Átsé Etsoh is associated with Átsé Ets' ózí in a relationship of complementarity. The two constellations cannot be seen in the night sky at the same time.

Scorpius

The constellation Scorpius includes two Navajo constellations, Átsé Etsoh and Gah Hahat'ee (Rabbit Tracks).

Many of the Greek stories of Scorpius are linked to battles with his adversary, Orion the Hunter. One story tells of Gaia, Mother Earth and creator of life, who became angry over Orion's repeated boasting that he could kill anything living on Earth. To silence Orion's bragging and to protect her creatures, Gaia sent the Scorpion to kill Orion.

Some stories say the giant scorpion did indeed kill Orion. Others say the rivalry never ended. In any case, both Orion and Scorpius reside in the sky in such a way that Scorpius and Orion are never seen at the same time (see the story of Orion on page 46).

Scorpius is considered an agent in the murder of several Greek figures. One of these was the son of Apollo, Phaethon. Apollo drove the Sun across the sky every day in a golden chariot. One day, Phaethon tried to drive the chariot by himself, but he was young and inexperienced. Along the way he saw the giant scorpion and became so frightened that he lost control of the horses. The chariot swerved wildly and the Sun fell too close to the Earth, burning lands and people. Phaethon himself was thrown from the chariot, tumbling into the stars of the great river Eridanus, where he was killed.

Apollo was saddened by the death of his son, but the lesson he learned was clear: the young man was inexperienced, did not have permission to drive the Sun in the golden chariot, and paid for his mistakes with his life.

Other stories mention the Scorpion's attacks on the Great Bull, Taurus. Many of these Greek stories have to do with the battle between Orion and Scorpius. The key players can often be observed clearly in the sky during late fall and early winter, when Scorpius goes down in the West and Orion and Taurus rise in the East.

As the Sun enters the constellation of Scorpius in November, it becomes increasingly weak. The days become shorter, and the heat of the Sun diminishes with the coming of winter. The Greeks saw this natural phenomenon as a continuous battle against the forces of darkness, played out as Orion and Taurus fighting Scorpius for control of the sky. At the winter solstice, the forces of darkness seem to have won a victory, but just at that moment, the Sun begins to regain its strength. The battle continues until the spring equinox, when the Sun passes into the constellation of Taurus and the days get longer than the nights once again.

GAH HAHAT'EE

NAVAJO NAME: Gah Hahat'ee
PRONUNCIATION: Gaa ha haat ay
TRANSLATION: Rabbit Tracks

WESTERN NAME: The Scorpion
LOCATION: Lower curved hook of Scorpius
OBSERVED PATH: East to West

GAH HAHAT'EE

The constellation Gah Hahat'ee is called Rabbit Tracks, because the four stars that outline it in the sky look exactly like the tracks that a rabbit would leave in the snow or sand. The name Gah Hahat'ee, however, refers to the movement a rabbit makes when it leaves its tracks—a simultaneous jumping up and leaping forward. This movement is inherent in the Navajo term.

Long ago people depended on rabbit meat and venison to survive. Rabbits were often hunted communally and provided an important part of the food supply. It has been said that rabbits were fatter in the winter because during that time they nibbled on the frost, in addition to the summer and fall vegetation and harvests.

The venison meat from deer was also an important staple in the traditional diet. The highest respect was accorded to the deer, and the hunt was carried out with cultural relevance and care. The reproductive cycle of the deer was considered significant. Hunting was carried out within the limitations of the reproductive cycle, so the deer population could continue its cycle of regeneration.

Gah Hahat'ee is located on the curved tail of Scorpius and travels in the southern sky from East to West. As late summer progresses into fall the constellation tilts toward the Western horizon, signifying that deer hunting can begin and indicating that the fawns have matured enough to survive on their own.

Both Navajo constellations Átsé Etsoh and Gah Hahat'ee are roughly located in Scorpius, one in the upper half and the other in the lower half. This is an example of how different constellations from different cultures don't always correlate exactly. The Greek comparisons for Gah Hahat'ee are the same as those already described for Átsé Etsoh.

CONSTELLATION COMPARISON

Hubble: Artist's rendition of a Black Hole near Scorpius.

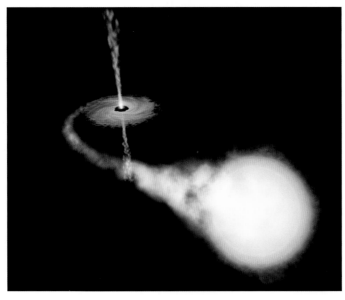

Often, many different constellations and cosmic objects are found in the same area of the sky. In the area of the constellation Scorpius, scientists have discovered a black hole.

Some astronomers say that stellar-mass black holes are made in supernova explosions. The companion star (orange globe) in the picture to the left apparently survived the original supernova explosion that created the black hole. It is an aging star that completes an orbit around the black hole every 2.6 days and is being slowly devoured by the black hole. Blowtorch-like jets (shown in blue) are streaming away from the black-hole system at 90 percent of the speed of light.

YIKÁÍSDÁHÁ / THE MILKY WAY

NAVAJO NAME: Yikáísdáhá
PRONUNCIATION: Yih kais daahaa
TRANSLATION: That Which Awaits the Dawn

WESTERN NAME: The Milky Way
LOCATION: The Milky Way
OBSERVED PATH: East to West

YIKÁÍSDÁHÁ

To the Navajo, Yikáísdáhá in its entirety is considered to be a constellation. The name Yikáísdáhá refers to the movement and different positions of the Milky Way throughout the night and through the various seasons, as they relate to the dawn. The name Yikáísdáhá can be translated as "that which awaits the dawn." To understand this term one should think in a traditional Diné cultural context. The Milky Way will be in a certain place in the night sky during the period just before dawn. The actual position will change

with the seasons and the nightly progression of the stars. This serves to illustrate the importance of the predawn night sky and the stars associated with this time of night.

There is only one time within the yearly cycle when the Milky Way will align with the entire predawn horizon. In fact the dawn and Yikáísdáhá are intrinsically connected as one. At this time Yikáísdáhá can be seen encircling the entire horizon in the predawn hours in the month of January, from Diné Bikéyah.

Yikáísdáhá is often depicted in traditional Navajo sandpaintings with an intertwined, zigzag pattern, reflecting its different movements and positions during the year. The patterns in the paintings illustrate how at different times of the year, the many stars of Yikáísdáhá appear to move from Northeast to Northwest, Northwest to Northeast, and even sit on the horizon.

THE MILKY WAY

The Greeks called the Milky Way *Galaxias Kuklos* or the Milky Circle. The Western word "galaxy" comes from the Greek *galaxias*. The Romans who came later called it *Via Lactea* or Milky Way. Many cultures have seen the Milky Way as drops of milk, or ashes, or a river.

The Greek story of the Milky Way begins with Zeus, king of all the gods, as so many of the stories do. Zeus was married to his Queen, Hera, but he had a fondness for many pretty young girls, and subsequently had many children. Most of his children remained mortals, not becoming gods. The only way for Zeus' children to become immortal was to drink milk from the bosom of Hera. This was, of course, very difficult to do, since Hera had a fierce jealousy of all of her husband's illegitimate children.

One story tells of how the strong warrior Hercules was an illegitimate child born to Zeus and a beautiful young woman named Alcmene. The gods admired the baby Hercules and wanted him to become immortal like them. So they conceived a plot where Hermes, the winged messenger, would place the baby upon the bosom of Hera while she was asleep. The baby drank a few drops of milk before Hera woke up, which made him an immortal god. Hera was so angry at this deception that she threw the baby off her, and drops of milk spurted up to the sky. These drops of milk became the Milky Way.

Another ancient story from the Middle East tells of the time just after the Great Flood. Noah's Ark was stranded on top of Mt. Ararat when the waters began to recede. At the top of the mountain, the survivors built an altar and had a ceremony to give thanks for their survival. The smoke from the fire went up into the heavens and became the Milky Way.

Other cultures see the Milky Way as a pathway for departed souls. This is true of many American Indians including the Iroquois, Algonquian, and some Plains Indians. To the Hindus of India, the Milky Way was a reflection of their sacred river, the Ganges. Many peoples in South America, including the Incas, identified figures of constellations in the dark spaces of the Milky Way, including the fox, the llama, the baby llama, the partridge, the toad, and the snake. Their stories interrelate the dark spaces, the stars, and the life cycles of these animals.

II'NI / PEGASUS AND LEO

NAVAJO NAME: Ii'ni
PRONUNCIATION: Ee neeh
TRANSLATION: Thunder

WESTERN NAME: The Winged Horse
LOCATION: Includes parts of Pegasus and Leo
OBSERVED PATH: East to West

II'NI

In Navajo, Ii'ni means thunder, but the constellation is often depicted as a Thunderbird. In traditional stories, the thunder is often described in the form of a bird. It is often paired with the energy of lightning. The spiritual essence of the constellation is visualized as a feather and is composed of six stars.

Ii'ni the Thunderbird is a most unique constellation, stretching across time and space, in a manner quite different from most Western constellations. The Thunderbird has a large square body preceded by a feather containing six stars. The large square body is the Greek constellation Pegasus and the six stars of the feather begin with Denebola in Leo, as observed in late September or early October, depending on the year.

The Thunderbird constellation is related to the Bear constellation (Shash), located in the Greek constellation Sagittarius, and both the Thunderbird and the Bear carry a relationship to the sound of the first thunder that occurs in the spring in Navajoland. The connection is made when both the complete feather of the Thunderbird and the nose of the Bear are visible in late February or early March. It is significant that the months must be the Navajo months, based on aspects of the lunar cycle, instead of the Western calendar months.

Much of the Thunderbird saga takes place within the parameters of the heliacal rise: a natural cosmic phenomenon that occurs when a star is first visible on the eastern horizon in the predawn morning sky. In Navajo cosmology, the emergence of predawn, well before sunrise, heralds the approach of significant stars, in this case the feather stars of Ii'ni, the Thunderbird. It may be good to note that Ii'ni, the Thunderbird, is really an essence or an energy, that relates to the sound made when lightning occurs, i.e., thunder.

In Navajo astronomy the feather of a star group (be ets'os) is highly significant, as it is a spiritual essence of a constellation, its connection to the entire cosmos. The first star of the Thunderbird's feather is a bright star that becomes visible in late September or early October, located in the East, with a heliacal rise at the time between the thin new crescent moon (dahit'á) and the full moon (hanibą́ąz). This star is Denebola, found at the tail of Leo the Lion. As the months proceed toward spring, each month features a new star, until the feather stretches out across the entire visible morning sky. In Navajo one could say, Ii'ni be ets'os ghaají biyiiyi silahagi.

The second star, Arcturus in Boötes, becomes visible in early November. Ii'ni be ets'os niłchi tsósí biyiiji silahagi. The third star becomes visible in early December in the east. Ii'ni be ets'os niłchitsoh biyii'ji silahagi. The fourth star, Vega in Lyra, becomes visible in late December or early January, depending on the year. Ii'ni be ets'os yasnilt'ees biyii'yi silahagi. The fifth star, Altair in Aquila, becomes visible in late January. Ii'ni be ets'os atsa biyáázh biyii'yi silahagi. The sixth star, Beta Pegasus, becomes visible in February, on the eastern horizon. Ii'ni be ets'os wóózhch'ííd biyii'yi silahagi. At the time of its heliacal rise, the nose of the bear constellation also becomes visible (star in Sagittarius).

In late February the first thunder occurs across the land of the Navajo. Navajos believe that the thunder and lightning cause the awakening and beginning of the annual vegetational growth. The thunder and lightning also awake the hibernating bears, reptiles, and insects. This cosmic event, which ties together the earth and the sky worlds, signals the completion of winter and the beginning of spring cosmic energies and activities. This is the end of Navajo Winter Stories for the year. In accordance with Navajo cultural protocol, from here until the coming of fall, one does not talk much about the sky world.

The first thunder varies by location. It is not tied to mechanical timekeeping devices. It may fluctuate somewhat from year to year. On the Navajo Nation, radio stations usually announce that the first thunder has come to this or that part of the reservation.

In the fall, insects, reptiles, and mammals, such as spiders, snakes, and bears, are often making preparations to slow down and hibernate. Plants are going underground or into a latent phase to gather their energies for spring renewal. It is at this time, as the earthly entities slow down and go underground, that the sky world introduces the Navajo constellation that will herald spring regeneration a half-year later. This is a striking illustration of Navajo concepts of relationship and complementarity.

The magnitude of the Navajo Thunderbird constellation is very great, but its true significance lies in the celestial relationships and multi-month process linked to cycles of the Sun, the Moon, and earthly regeneration. In comparison to other cultures, it is interesting to note that the ancient Chinese identified some of the stars of the Western constellation, Lacerta, as a cosmic serpent which awoke at the end of winter in late January. To the ancient Chinese the serpent was visible when the stars of Lacerta appeared in the south, around midnight. This awakening is similar to the Navajo concept of reptiles awakening and emerging at the beginning of spring, in connection with specific star alignments.

PEGASUS

When you look up at the sky, the Greek constellation of Pegasus is best recognized as a square with a specific star at each corner. The horse has two large wings and curiously, it always looks upside down in the sky. The stars of Pegasus show only the front half of a horse, with the great square composing the body.

Many ancient people besides the Greeks saw this constellation as a horse. Ancient ships of the Mediterranean Sea often had figureheads of a winged horse on their bows to provide navigational guidance. In the time of the Sumerians, in the area now called Iraq, people called the stars in Pegasus "The Center of the Universe" and also thought of them as the location of paradise.

The ancient Greeks connected the winged horse to Poseidon, the great God of the Sea. Some Greek stories also say Pegasus was the son of Poseidon.

Pegasus is related to water in many ways. Some versions of the stories say he is the creator of fountains and springs. It is said that one day, Pegasus kicked the earth on Mt. Helicon and caused a fountain of water to spring out. Pegasus is connected to ceremonies designed to bring rain.

The birth of Pegasus goes back to the story of Perseus, the warrior who killed Medusa, a princess who had been turned into an extremely ugly woman with hair of snakes and the body of a bird. Perseus chopped off Medusa's head and a beautiful horse jumped out of what had been her severed neck. This horse was Pegasus.

Pegasus grew up on Mt. Helicon and was much loved by the young Muses. He lived wild and free until he was tamed by Bellerophon. Pegasus and Bellerophon had many adventures together.

As Bellerophon became a famous warrior, he also became arrogant and vain. One day he decided to attempt the dangerous journey to Mt. Olympus, the home of the Greek gods. This was not looked on with favor by the gods, since Bellerophon was not an immortal himself.

Zeus decided to prevent Bellerophon from reaching Mt. Olympus and sent

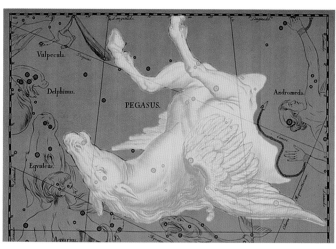

a horsefly to bite Pegasus. The horsefly bit Pegasus just under the tail, causing him to buck and throw Bellerophon off into a bush of thorns. Bellerophon was crippled and blinded by the fall and remained so for the rest of his life.

Pegasus, however, was warmly welcomed into the kingdom of the gods at Mt. Olympus by Zeus himself.

He lived in the royal stables and was given the responsibility of carrying the great bolts of lightning made by the one-eyed Cyclops. When

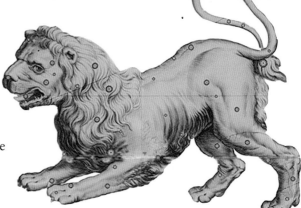

the constellation of Pegasus first becomes visible in the predawn hours of early spring, thunder and lightning occur. It is interesting to note that the Greeks associated Pegasus with lightning and the Navajos associate the same stars of Pegasus with Ii'ni, the Thunder.

LEO

The first star of the feather of the Navajo Thunderbird, Ii'ni, is Denebola, the tail of Leo the Lion. The constellation of Leo the Lion usually is associated with sovereignty, heroes, and the Sun. The Greeks say that Leo was created by Selene, the Moon, from sea foam. The lion was killed by Hercules, who from then on wore a lion skin.

Lions were very important in the Mediterranean world. In ancient Eygpt, lions would come to the Nile River during the flooding period. This usually occurred around the time of the summer solstice, which was also the time of the heliacal rise of the star Sirius.

The Egyptians said the arrival of the lions signaled the beginning of the New Year. The relationship of lions and water was illustrated by the many fountains that were created showing the head of a lion, with water flowing out of its mouth. Even the famous Sphinx has the look of a lion.

Many Mediterranean stories were associated with the power of Leo during August, when the Sun travels through the constellation. Some say that electrical storms can spoil milk and disturb one's blood during this time. It was also thought to be dangerous to take a bath in August, due to the electrical influences from Leo and the Sun.

CONSTELLATION CONNECTION

As mentioned before, Navajo and Western constellations cannot always be directly correlated. In the case of Ii'ni, it uses one star from Leo, parts of Pegasus, and several other constellations. The Western constellations can be useful reference points in the night sky when you are first trying to see some of the Navajo constellations. But seeing Navajo constellations requires a different concept of time and space. For example, the constellation Ii'ni actually takes about nine months to fully emerge in the sky.

Off the northeast corner of the great square of Pegasus lies the Andromeda Galaxy—a collection of hundreds of billions of stars.

The image below left was taken by the NASA Galaxy Evolution Explorer (GALEX) satellite in ultraviolet light. The blue wisps are areas of hot, young stars. If you could get up close, they might look like the hot, young stars you see in the Pleiades star cluster (page 42) in our own galaxy. The yellowish-white area in the center of the galaxy is an area of older, cooler stars.

TOP: Hubble: Andromeda.

BOTTOM: Hubble: Virgo Cluster.

The Andromeda Galaxy is the most distant object that you can see without the aid of binoculars or telescopes. It lies 2.5 million light years away from Earth. The light from it you are seeing now left the Andromeda Galaxy 2.5 million years ago. You are looking into the past, seeing what it looked like before the first human beings appeared on Earth.

Just to the east of Denebola, the tip of Leo's tail, is the Virgo Cluster of Galaxies. This ultraviolet light image from the NASA GALEX satellite shows just a small portion of the cluster. As seen from Earth, the entire Virgo Cluster covers an area of the sky 10 times larger than the full moon. Measured from edge to edge, the cluster spans some 15 million light years of space.

There are perhaps 2,000 galaxies in this cluster, with each galaxy containing countless billions of stars. Spiral galaxies appear as elongated structures with blue areas of young stars. Elliptical galaxies appear as round yellow/orange/red structures that contain mostly older stars.

The Virgo Cluster is the closest cluster of galaxies to us, yet it lies about 60 million light-years away. The light we see now left the Virgo Cluster 60 million years ago, when the Rocky Mountains were still rising to their current height.

Resembling the fury of a raging sea, this image actually shows a bubbly ocean of glowing hydrogen gas and small amounts of other elements such as oxygen and sulfur.

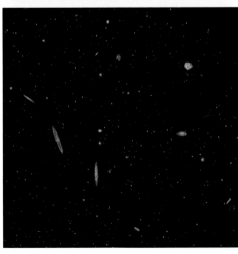

The wave-like patterns of gas have been sculpted and illuminated by a torrent of ultraviolet radiation from young, massive stars, which lie outside the photo to the upper left. The glow of these patterns accentuates the three-dimensional structure of the gases. The ultraviolet radiation is carving and heating the surfaces of cold hydrogen gas clouds. The warmed surfaces glow orange and red in this photograph. The intense heat and pressure cause some material to stream away from those surfaces, creating the glowing veil of even hotter greenish gas that masks background structures. The pressure on the tips of the waves may trigger new star formation within them.

This photograph, taken by NASA's Hubble Space Telescope, captures a small region within M17, a hotbed of star formation. M17, also known as the Omega or Swan Nebula, is located about 5,500 light-years away in the constellation Sagittarius. The image, roughly three light-years across, was taken May 29 to 30, 1999, with the Wide Field Planetary Camera 2. The colors in the image represent various gases. Red represents sulfur; green, hydrogen; and blue, oxygen.

Hubble captures a perfect storm of turbulent gases.

SHASH / SAGITTARIUS

NAVAJO NAME: Shash
PRONUNCIATION: Shaash
TRANSLATION: Bear

WESTERN NAME: The Archer
LOCATION: Approximately Sagittarius
OBSERVED PATH: East to West

SHASH

The constellation Shash, along with Ii'ni, heralds the coming of spring and summer. Shash appears in the predawn hours in early spring when its nose rises above the horizon, at the same time that the body of Ii'ni begins to appear.

This is when the First Thunders are heard in early spring, which awakens the hibernating bears, reptiles, and insects, as well as the dormant plants.

The natural cycles of earth and sky are united by the seasonal processes of the Ii'ni and Shash constellations. The union of earth and sky—often referred to as Mother Earth and Father Sky—is experienced through the spring, summer, fall, and winter seasons. With the coming of spring, for example, Ii'ni (Thunder) and rain occur along with winds and warmer air. As the earth begins to warm up, certain plants begin to grow, heralding the approach of spring and summer.

The plants that grow are intrinsically connected to the animals that eat them, and this is often reflected in the names of the animals. For example, Shash daa means bear food such as berries, and dahiitíí daa means hummingbird food.

SAGITTARIUS

The ancient Greeks saw the constellation of Sagittarius as a hunter, specifically an archer with a drawn bow and arrow ready to shoot at Scorpius, his enemy. The archer is not wholly human, however. He has the upper body and head of a man but the lower part of his body is a horse.

Sagittarius was raised on a mountain with his foster sisters, the Muses. The Muses are the goddesses of various arts, such as music and dance. Even today, in English, we may refer to the Muses as inspirations for the arts.

Sagittarius became the close companion of the Muses and greatly admired their talent. It is said that he invented applause, to show his admiration for their talents. The Muses in turn loved and admired their foster brother, Sagittarius. They begged Zeus to place him in the sky as a reward for his skills and friendship. Zeus agreed and Sagittarius joined the other heroes in the celestial kingdom.

Sagittarius is also known as the guardian of one of the doors of reincarnation. When a Greek citizen would die, he or she had to pass through two special doors on the way to reincarnation, before returning to life on Earth. The doors were near the Milky Way, which was considered a resting place for souls after death. The souls had to wait there until the fall equinox. When this occurred, souls could then pass through the first door of Sagittarius, and then Gemini, on their way to reincarnation.

TRIFID NEBULA

This NASA Hubble Space Telescope image of the Trifid Nebula reveals a stellar nursery being torn apart by radiation from a nearby, massive star. The photo also provides a peek at embryonic stars forming within an ill-fated cloud of dust and gas, which is destined to be eaten away by the glare from its massive neighbor. This stellar activity is a beautiful example of how the life cycles of stars like our Sun is intimately connected with their more powerful siblings.

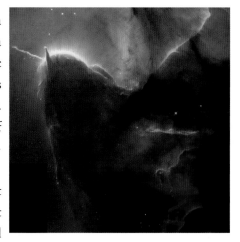

The Trifid Nebula.

The Hubble image shows a small part of a dense cloud of dust and gas, a stellar nursery full of embryonic stars. This cloud is about eight light-years away from the nebula's central star, which is beyond the top of this photo. Located about 9,000 light years from Earth, the Trifid resides in the constellation Sagittarius.

A stellar jet (the thin, wispy object pointing to the upper left) protrudes from the head of a dense cloud and extends three-quarters of a light year into the nebula. The jet's source is a very young stellar object that lies buried within the cloud. Jets such as this are the exhaust gases of star formation. Radiation from the massive star at the center of the nebula is making the gas in the jet glow, just as it causes the rest of the nebula to glow.

TSETAH DIBÉ

NAVAJO NAME: Tsetah Dibé
PRONUNCIATION: Tseh tah dibeh
TRANSLATION: Mountain Sheep

WESTERN NAME: The Crab
LOCATION: Cancer
OBSERVED PATH: East to West

TSETAH DIBÉ

The Mountain Sheep is a winter constellation. It is a very small cluster of stars, much dimmer than Dilyéhé, the Pleiades. It is located in the Beehive Cluster of Cancer, somewhat overhead in Diné Bikéyah in December, about midnight.

Navajos have closely observed Tsetah Dibé at this time of year, as this constellation is a major timekeeper for certain healing ceremonies. One of these is the nine-night winter ceremony, Nightway (often referred to as Yei bi Chei), where certain ceremonial events occur throughout the night. The movement of this constellation, from east to west, is significant as a time indicator to tell the coming of the dawn, and thus the completion of parts of the Nightway ceremony.

The Mountain Sheep plays a prominent role in the Navajo culture, and the ancient cultural songs sung in the Nightway ceremony include references to the Mountain Sheep.

This constellation is most visible to the observer in December, when it is very cold outside and the Moon is not too bright. At this time, sound travels farther than normal, a phenomenon that was noticed by traditional people.

One of the scientific reasons one can see better on cold nights is the absence of particles and water vapor in the air. Particularly with snow on the ground and low temperatures, there is less vapor and dust in the sky to block and refract the light traveling from faraway stars.

CANCER

Cancer the Crab is most highly visible on dark moonless nights from December to spring. However, like Tsetah Dibé, it is difficult to see unless you are in a very dark place, on a cold night, when stars are sparkling.

There are several Greek stories about Cancer. One story says that Queen Hera placed Cancer in the skies because it pinched Hercules on the foot while he was fighting the water snake, Hydra, in the marshes.

Another Greek story refers to two of the stars in Cancer as little donkeys, who are separated by a nebula which was called the manger. The manger is in the stable where animals are housed. This manger is called Praesepe, and is known to astronomers as M44. Some people refer to the manger as a beehive, since it is a faint cluster of stars.

The Beehive Cluster in Cancer.

Yet another story links the two little donkeys with Queen Hera and Dionysus, the god of wine and drinking. Though the Queen was the wife of Zeus, Dionysus fell madly in love with her. She did not welcome his attention, and in fact greatly disliked Dionysus, because he was the illegitimate son of her husband Zeus and Selene the Moon.

The only way for Dionysus to overcome his lovesickness was to visit the Oracle of Zeus, at Dodona. He traveled for a long time searching for the oracle. Near the end of his journey, he came to a wet marshy area that seemed impassable. Two little donkeys carried him and his traveling companion across the marsh to the oracle, where Dionysus was cured of his lovesickness. He was so grateful to the donkeys that he made them immortal and placed them in the sky.

Another story also involved Zeus and the donkeys. Zeus was angry at the Titans and declared war on them. He called for all the other Greek gods to assist him in the battle. The gods came to his aid, as requested. Dionysus and his army arrived riding on small donkeys. When the army drew close to the Titans, the donkeys were overcome with fear and brayed loudly. The Titans were so frightened by the noisy braying that they turned and ran, leaving Zeus victorious in the battle with the Titans. Zeus was so pleased with the donkeys that he turned two of them into immortals and placed them into the sky.

Other cultures saw the crab constellation as different animals. In Mesopotamia, in the Middle East, it was thought to be a turtle. On the island of Crete in the Mediterranean Sea, the constellation was thought to be an octopus. In ancient Egypt, it was thought to be a scarab beetle and a replication of the scarab was often placed over the heart of a mummy during funeral services.

Hubble: View of a
planetary debris disc.

Hubble: Artist's view
of planet and disc around
the star AU Microscopii
near Cancer.

AU MICROSCOPII

NAME: AU Microscopii
DESCRIPTION: Spectral Type "M0" Star
POSITION (J2000): R.A. 20h 45m 09s.53 Dec. -31° 20'27".2
DISTANCE: 33 light years (10 parsecs)
CONSTELLATION: Microscopium

Two of NASA's great observatories, the Spitzer Space Telescope and the Hubble Space Telescope, have provided astronomers an unprecedented look at dusty planetary debris around stars the size of our Sun.

Spitzer has discovered for the first time dusty discs around mature, Sun-like stars known to have planets. Hubble captured the most detailed image ever of a brighter disc circling a much younger Sun-like star. The findings offer "snapshots" of the process by which our own solar system evolved, from its dusty and chaotic beginnings to its more settled present-day state.

"Young stars have huge reservoirs of planet-building materials, while older ones have only leftover piles of rubble. Hubble saw the reservoirs and Spitzer, the rubble," said Charles Beichman, PhD, NASA Jet Propulsion Laboratory (JPL). He is lead author of the Spitzer study. "This demonstrates how the two telescopes complement each other," he added.

The young star observed by Hubble is 50 to 250 million years old. This is old enough to theoretically have gas planets, but young enough that rocky planets like Earth may still be forming. The six older stars studied by Spitzer average 4 billion years old, nearly the same age as the Sun. They are known to have gas planets, and rocky planets may also be present. Prior to the findings, rings of planetary debris, or "debris discs," around stars the size of the Sun had rarely been observed because they are fainter and more difficult to see than those around more massive stars.

Debris discs around older stars the same size and age as our Sun, including those hosting known planets, are even harder to detect. These discs are 10 to

100 times thinner than the ones around young stars. Spitzer's highly sensitive infrared detectors were able to sense their warm glow for the first time.

1. Why do stars have discs?

When a star forms through the collapse of a huge cloud of gas and dust, some of the material will settle into a flattened disc around the star that is at a right angle to the star's spin axis. This material has maximum centripetal force along this plane and so resists falling into the star. The disc eventually dissipates after the star ignites. But before it does, planets, asteroids, and comets may agglomerate from dust sticking together in the disc. A secondary disc forms when the smaller bodies collide and grind each other back down to dust.

2. Do all stars have debris discs?

NASA's Spitzer Space Telescope has detected the telltale infrared glow of excess dust around a number of young stars. However, some stars appear to lose material too early to make planets. What's new is that some of the stars with debris discs have planets, as identified in previous studies. These planets have never been directly seen but rather are identified through the telltale gravitational wobble they induce on the star. The new Spitzer observations are the first direct link between existing extrasolar planets and circumstellar discs.

3. Does our Sun have a debris disc?

Secondary discs around young stars are very dusty due to an outpouring of violent collisions. Our solar system is somewhat middle-aged, and the rate of collisions has died down dramatically. Still, the ecliptic plane of our solar system, where all the major planets and asteroids lie, has about one ten-thousandth as much dust as seen around young stars. This can be seen from Earth as a thin pillar of faint light in the night sky called the Zodiacal Light. It is best seen from the tropics where the ecliptic is typically at a steep vertical angle to the horizon.

4. Where can I find out more about this topic?

For more information from HubbleSite and beyond, visit the Related Links at http://hubblesite.org/newscenter/newsdesk/archive/releases/2004/33/related

TŁISH TSOH / CANIS MAJOR

NAVAJO NAME: Tłish Tsoh
PRONUNCIATION: Tli sh tso
TRANSLATION: Big Snake

WESTERN NAME: Great Dog
LOCATION: Parts of Puppis and Canis Major
OBSERVED PATH: East to West

TŁISH TSOH

The constellation Tłish Tsoh is located in the southern part of the night sky. It is a highly respected constellation that is associated with healing. With the coming of winter Tłish Tsoh becomes visible. Thus, Tłish Tsoh is an indicator that winter ceremonies can be performed and winter stories, including star stories, can be told.

Navajos say that when snakes are hibernating you can tell winter stories. In the land of the Navajo, Diné Bikéyah, snakes will go into the ground and enter into a period of winter seasonal hibernation, where their whole physical body will slow down and enter into a form of extended sleep. During this time they do not move around much or eat.

In contrast to the winter months, with the coming of spring Tłish Tsoh becomes invisible. During this time the First Thunder is heard, sending vibrations deep into the earth, waking up the snakes and other reptiles who have been hibernating. At this time they begin to emerge onto the earth surface.

Thus in the spring and summer when snakes are active on the earth surface, the Tłish Tsoh constellation is not visible in the night sky. But with the coming of winter the snakes go into the earth and hibernate, just as the Big Snake constellation becomes visible in the southern sky.

CANIS MAJOR

Canis Major is seen as a large dog in the astronomical traditions of both the Egyptians and Greeks. The Greeks located Canis

Major south of the feet of Orion, standing on his hind feet, ready to jump on Lepus the Rabbit.

A short distance to the north lies Canis Minor, the Little or Lesser Dog. The two dogs, Canis Major and Canis Minor, belonged to the hunter Orion. The concept of the dog days of summer may have originated with these celestial dogs, as they are identified with the 40 hottest days of late summer. The term "canine" comes from the word Canis, meaning dog.

A very bright star called Sirius is located near the nose of Canis Major. The name Sirius comes from the Greek word *seir* meaning "to shine." Sirius is the brightest star in our sky.

Canis Major and its brightest star Sirius acted as a weather indicator for the agro-astronomy culture of ancient Egypt. The Egyptian New Year began at the heliacal rise of Sirius during the time of the summer solstice. When Sirius was seen, Egyptians knew that hot winds would pass over the Nile River, bringing rain clouds and major rainstorms. The Nile River would subsequently flood over the fertile flood plains and the Egyptians would plant their crops. Sirius and Canis Major were eagerly anticipated by Egyptian farmers.

Hubble: spiral galaxies.

HELIACAL RISE

In Western astronomy, heliacal rise refers to a star's first appearance on the eastern horizon near sunrise. However, to traditional Navajos, heliacal rise refers to a brief span of time—about 20 minutes—that correlates with the first emergence of sunlight over the eastern horizon (háyílkááh) in the predawn, well before sunrise. Five minutes later the light will spread out to the north and south along the eastern horizon, at the same time rising slightly above the horizon (naniinílkááh). Another five minutes later, the light will spread farther north, south, and upward, but now it is brighter (dahidílkááh). The completion of the predawning process (ałso'yiskááh), according to traditional Navajos, is the appearance of a small streak of pinkish-red, just above the horizon, below the spreading light.

During the time of naniinílkááh and dahidílkááh the first appearance of the star or constellation that is contained within the predawn light is the Navajo heliacal rise of that star or constellation.

Traditional Navajo reliance on the predawn and heliacal rise of certain stars cannot be overestimated. This is a very important time indicator, in addition to serving as an identification of constellations associated with the Moon calendar. Predawn is also the demarcation for the beginning of a new day, in contrast to the Western association of midnight with the beginning of a new day.

Certain stars and constellations are identified with certain calendar months of the year. The Navajo months start with the appearance of the first crescent moon, and the stars or constellations that appear in the days just after the first crescent moon are the time indicators for the heliacal rise and beginning of new months.

In the direction of the constellation Canis Major, two spiral galaxies pass by each other like majestic ships in the night. The near-collision has been caught in images taken by NASA's Hubble Space Telescope and its Wide Field Planetary Camera (see photo on page 71).

Long streamers of stars and gas stretch out a hundred thousand light years toward the right-hand edge of the image.

MA'II BIZQ' / ARGO NAVIS

NAVAJO NAME: Ma'ii Bizq'
PRONUNCIATION: Mah ee bihzon
TRANSLATION: Coyote Star

WESTERN NAME: The Fast Ship
LOCATION: Canopus (in Argo Navis)
OBSERVED PATH: East to West

MA'II BIZQ'

Canopus, called the Coyote Star in Navajo, is one of the brightest stars in the sky, second only to Sirius. It is also called the Monthless Star in Navajo because it is visible in the southern sky for less than a month. Its color, as seen from Diné Bikéyah, is usually red or bright orange, and it is often twinkling. From Diné Bikéyah it appears to move in a small, semicircular orbit on the southern horizon.

It is said that Coyote, the Trickster, wanted to participate in the naming and placing of the stars and constellations at the beginning of time. According to Navajo cosmogony, Coyote impatiently grabbed the buckskin pouch containing the crystal seeds of the stars and placed one star in the far south, claiming it for himself. He then tossed all the remaining crystals high into the sky, providing chaos and disrupting the careful order that was envisioned. Navajos say it is because of this chaos that many stars exist with no names.

For the complete story, listen to the CD "Stars over Diné Bikéyah," available from the website www.sharingtheskies.com.

ARGO NAVIS

The constellation Argo Navis represents a large ship. *Argo* means "fast" and *Navis* refers to "ship." Most of Argo Navis is contained in the Milky Way and only one star shines out brilliantly, the multi-colored star Canopus, which in fact is the second-brightest star in the sky.

The Argo Navis was such a large constellation that French astronomers in the eighteenth century divided it into three smaller constellations: Carina (the keel), Puppis (the stern or back part of a ship), and Vela (the sails). Today, Puppis depicts the stern end of the ship while the front part of the ship is missing on most of our star maps. Stars from Puppis are also part of the Navajo Tłish Tsoh constellation, along with stars of Canis Major.

The Argo Navis constellation is a special protector for fishermen and sailors. Some say it guided Noah in his ark during the Great Flood. Many other cultures, including the Egyptians and the early Hindus, refer to this constellation as a ship, with Canopus as its pilot and navigator.

Many Greek stories about Argo Navis exist. Some say it was the first ship to sail the seas. Others say that Chiron the Centaur (half man and half beast) sailed Argo Navis. During the voyage, Chiron designed a globe with 48 constellations as a guide for the Argonauts. In later years, Christian astronomers called Argo Navis the Ark of Noah.

One Greek story says the ship Argo was built by Jason, a Greek hero, to avenge the loss of his father's land. This is the basis of the famous Jason of the Golden Fleece stories, which tell of a fantastic set of adventures that occur while Jason and the Argonauts travel the Mediterranean, searching for the golden sheepskin. Jason had to travel to the forests of Hades and retrieve the Golden Fleece so he could recapture his family's lost homeland.

Jason commissioned the Argo to be built and gathered 50 Greek heroes to accompany him on his travels. Since these warriors were sailing on the ship Argo, they were called the Argonauts. The Argonauts had many adventures and in the end they did succeed in capturing the Golden Fleece.

Many of our contemporary English words do in fact come from ancient Greek. Present-day space voyagers are called astronauts (star sailors), a very appropriate name. A contemporary use of the word "navis" is our word "navy" for the nautical part of the U.S. Armed Forces.

The bright star in Argo Navis, Canopus, appears multi-colored, sometimes blue and sometimes red. It can be seen twinkling due south and very low on the horizon at certain times of the year in Navajoland. In more southern locations such as Hawaii, you can see it all year long and much higher in the southern sky.

Canopus was named after King Menelaus' chief warship navigator, Kanobus. The Greek warriors were returning home after the end of the Trojan War after defeating Troy. This was the time of the Trojan Horse, when legend says the Greeks secretly hid many of their soldiers inside a great wooden horse and wheeled it up to the gates of Troy. The Trojans thought the horse was a gift and brought it inside their city walls. During the night the soldiers crept out of the hollow horse and captured the city.

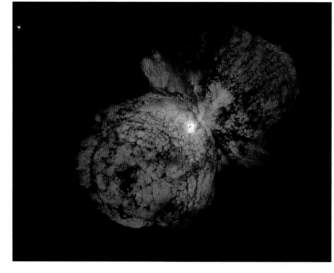

Hubble: Exploding, super-massive star in Argo Navis.

On the way home from Troy, King Menelaus and his navigator Kanobus stopped in Egypt and there Kanobus was bitten by a poisonous snake and died. The king was devastated by the loss of his chief navigator and built a monument to him, which later grew into the city of Canopus. The king also named the bright southern star after his navigator. Thus the star Canopus still guides the Argo Navis in the celestial sky.

Desert nomads saw Canopus as a camel, al-Fahl. They said the blue shine from Canopus gave brilliance to jewels and also protected them from sickness. Even today, desert nomads predict the end of the hottest weather of the summer through the heliacal rise of Canopus.

EXPLODING STAR

A huge, billowing pair of gas and dust clouds is captured in the stunning Hubble telescope photo (page 74) of the super-massive star Eta Carinae, inside Argo Navis.

Eta Carinae is more than 8,000 light-years away and 10 billion miles across (about the diameter of our solar system). Eta Carinae suffered a giant outburst about 150 years ago, when it became one of the brightest stars in the southern sky. Though the star released as much visible light as a supernova explosion, it survived the outburst. Somehow, the explosion produced two lobes and a large, thin equatorial disc, all moving outward at about 1.5 million miles per hour. Estimated to weigh 100 times more than our Sun, Eta Carinae may be one of the most massive stars in our galaxy.

EAGLE NEBULA, M16

NAME: Eagle Nebula, M16
DESCRIPTION: Nebula
POSITION (J2000): R.A. 18h 18m 51.06s Dec. -13° 49'51.11"
DISTANCE: 6,500 light years
CONSTELLATION: Serpens

Undersea corral? Enchanted castles? Space serpents? The eerie, dark pillar-like structures on page 76 are actually columns of cool interstellar hydrogen gas and dust that are also incubators for new stars. The pillars protrude from the interior wall of a dark molecular cloud like stalagmites from the floor of a cavern. They are part of the Eagle Nebula, a nearby star-forming region 6,500 light-years away.

The pillars are in some ways like buttes in the desert, where basalt and other dense rock have protected a region from erosion, while the surrounding landscape has been worn away over millennia. In this celestial case, it is especially dense clouds of molecular hydrogen gas and dust that have survived longer than their surroundings in the face of a flood of ultraviolet light from hot, massive newborn stars. This process is called "photoevaporation."

Pillars of creation in a star-forming nebula.

Ultraviolet light is also responsible for illuminating the convoluted surfaces of the columns and the ghostly streamers of gas boiling away from their surfaces, producing the dramatic visual effects that highlight the three-dimensional nature of the clouds. The tallest pillar (left) is about four light-years long from base to tip.

As the pillars are slowly eroded away by the ultraviolet light, small globules of even denser gas buried within the pillars are uncovered. These globules have been dubbed "EGGs," or Evaporating Gaseous Globules, but it is also a word that describes what these objects are. Forming inside at least some of the EGGs are embryonic stars. Eventually, the stars emerge as the EGGs succumb to photoevaporation.

This photo was taken by the Hubble Space Telescope Wide Field and Planetary Camera 2. The color image is constructed from three separate images taken in the light of emission from different types of atoms.

JO'HANAA'ÉÍ / THE SUN

NAVAJO NAME: Jo'hanaa'éí
PRONUNCIATION: Joe haan aa ay
TRANSLATION: Sun

WESTERN NAME: Sun
LOCATION: Sun
OBSERVED PATH: East to West

JO'HANAA'ÉÍ

It is generally understood by most indigenous cultures, including the Navajo, that the Sun provides light and heat and is a source of energy for life. To the

Navajo, the Sun provides an adequate amount of light and heat for the growth of life on Earth. The division between day and night throughout the year enables natural life processes and dynamic growth, as well as providing a shield and protection from the direct and potentially harmful radiation of the Sun. Too much Sun would scorch and burn the Earth. Too little Sun would diminish potential for life on Earth.

Like many Navajo constellations, the Sun and the Moon are generally thought of as a complementary pair, partially because Sun and Moon appear to be of similar size when viewed from Diné Bikéyah. The Sun is said to be the male energy of the pair and the Moon is said to be the female energy. Together their cycles determine time: hours, days, months, and years.

There are numerous Navajo stories about the Sun. Some talk about the Sun as a great turquoise disc that is carried by a being, sometimes depicted as an old man, who rides a horse across the sky each day.

Other stories talk about the Sun as the father of Monster Slayer, one of the Navajo hero twins, who went to the Sun with his brother, Born for Water, to get weapons with which he could slay the monsters who were making life miserable on Earth. The Sun gave the Twins many tests to make sure they were really his sons and that they were strong enough for the tasks ahead. The Twins proved themselves worthy and were given the weapons. With the weapons, they returned to Earth to slay many monsters, making life safe for humans.

Still other stories talk about Changing Woman as a spiritual companion of the Sun. It is said that the Sun built a beautiful house for her in the waters of the West. She lived there for a long time and created children from her body who became the foundation of the Navajo clan system.

The Sun is an important timekeeper for the Navajo. Its pathway is acknowledged with the winter and summer solstices, along with the four or so days where it appears to stand still, at the time of the solstices. The spring and fall equinoxes are equally important as these are the times when the Sun is rising due east. During an equinox, the day and the night are of equal duration, each lasting twelve hours.

For the Navajo, the cardinal directions are indicated by the positions of the Sun: East in Navajo is Ha'a'aah, meaning, "Where the Sun comes up." East is where the Sun rises at the time of the equinox.
South is Shádiaah, which means, "The Sun travels with and for me."
West is Ee'ee'aah, "Where the Sun goes down."
North is Náhookǫs, and relates to the Náhookǫs constellation.

As opposed to Western cardinal directions, where north is put at the top of maps, east is at the top for the Navajos. Traditional Navajo sandpaintings, as well as most ceremonies and stories, begin in the East.

The dome-like structure of the Navajo hogan demarcates the cardinal directions and cycles of the Sun for the solstice and equinox. Since the doorway of a traditional hogan faces east, on the day of the equinox the Sun enters through the door and shines on the back wall. Most of the year, the path of the Sun can be traced on the back wall of the hogan. You can actually use the back wall of the hogan as a calendar.

The time of predawn is very important in Navajo timekeeping. The Navajo language has very precise terminology for each stage of the dawning process, which, to the Navajos, precedes the rising of the Sun. It is during this time that certain stars become visible. This is called heliacal rise. The heliacal risings of certain stars serve as important markers for timekeeping, planting ceremonies, and for signifying the end of winter and the beginning of spring.

Navajos say that the hottest days of the year come two to three weeks after the summer solstice. Similarly, the coldest days of the year come two to three weeks after the winter solstice.

The shadow that is made by the Sun is very important in Navajo culture. It is said that one is born with a shadow and that the shadow stays with one until life comes to an end.

THE SUN: HELIOS

The Sun is so critical to life on Earth that it has been studied and honored throughout the ages. Almost universally, the Sun represents the day and the

Moon represents the night. Both the day and the night are necessary for the process of natural order.

Most Western traditions name the cardinal directions east and west from the daily path of the Sun from dawn to dusk. North and south are derived from the yearly path of the Sun from solstice to solstice.

Many people say that the Sun is carried across the sky. The Greeks said the Sun was driven in a golden chariot, by Apollo. The Navajo said the Sun was carried by an old man on a horse. The Egyptians said the Sun was carried in a sailboat.

The Greek stories talk about Eos, the Dawn, the granddaughter of earth and sky. Eos was married to Astraeus, the Starry One, who was the God of the Night Sky. Astraeus had several children, including the Four Winds of the cardinal directions, and Astraeus and Eos were the parents of the Morning Star. The brother of Eos was Helios, the Sun, and the sister of Eos was Selene, the Moon.

The Greeks said that every morning Eos, the Dawn, drove her own chariot across the sky, parting the gates of the sky for her brother, Helios, the Sun, who was following her in his chariot.

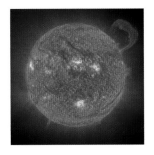

Handle-shaped prominence erupting off the Sun.

PROMINENCES

Prominences are huge clouds of relatively cool dense plasma suspended in the Sun's hot, thin corona. At times, they can erupt, escaping the Sun's atmosphere. Emission in this spectral line shows the upper chromosphere at a temperature of about 60,000 degrees K. Every feature in the image traces magnetic field structure. The hottest areas appear almost white, while the darker red areas indicate cooler temperatures.

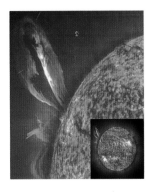

Large erupting prominence—Earth shown for size comparison.

SOLAR STORMS

Once thought to be unchanging, the Sun is now known to vary constantly. Changes in the activity of the Sun occur in 11-year cycles. Sunspots can appear and disappear over days or weeks. Flares and large ejections of mass (coronal mass ejections) occur in time spans of minutes to hours. The energy of the Sun constantly blows out a "solar wind" of electrified particles that is the extended atmosphere of the Sun.

Abrupt changes on the Sun can create flares and coronal mass ejections that blast brief but powerful "solar storms" into space. Earth is surrounded by a

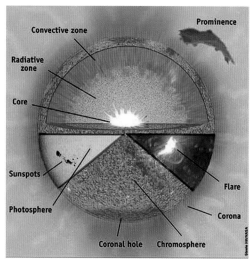

Prominence

Convective zone

Radiative zone

Core

Sunspots

Photosphere

Coronal hole   Chromosphere

Flare

Corona

The parts of the Sun.

magnetic field (magnetosphere) that protects us from the worst effects of solar storms. However, solar storms can cause fluctuations in the magnetosphere called magnetic storms. These magnetic storms have disabled satellites and burned out transformers, shutting down power grids. These storms also can endanger astronauts. The storms contribute to more intense auroras that can be seen closer to the equator than is usual.

THE AMAZING STRUCTURE OF THE SUN

The diameter of the Sun is about 1,390,000 km. The average diameter of the Earth is 12,740 km. About 109 Earths could be placed side-by-side along the diameter of the Sun.

The volume of the Sun is unimaginable. Approximately 1,300,000 Earths could fit inside the Sun. Its mass is about 300,000 Earths.

The structure of the Sun is separated into several regions: Interior, Photosphere, Chromosphere, Transition Region, Corona, and Solar Wind.

THE INTERIOR: THE CORE AND THE RADIATIVE LAYER The core occupies the first 25 percent of the distance from the center. The temperatures range between 15,000,000°C at the center to 7,000,000°C at the outer edge of the core. Most metal is hot enough to glow white at only about 6,000°C. The density changes dramatically. The Sun is eight times the density of gold at the center. At the outer edge of the core, the density is about the same density of gold. At these temperatures and pressure the hydrogen is fused into helium.

Surrounding the core is the radiative zone. The radiation that escapes the core is mostly x-rays. This radiation takes about 1 million years to find its way out of the radiative layer, even traveling at the speed of light, due to collisions between light and matter within the radiative layer.

Between the radiative zone and the convective zone is an interface layer. It is now believed that there is a magnetic dynamo in this layer that generates the Sun's magnetic field.

In the final 200,000 km to the photosphere, energy is carried by convection in the convective zone. The temperature drops from 2,000,000°C to 5,700°C.

The density drops from 0.2 g/cm³ to 0.0000002 g/cm³. Hot plasma rises and cooler plasma sinks, creating "cyclones" as the Sun rotates.

THE PHOTOSPHERE As these bubbles of upwelling, hot plasma reach the surface of the photosphere, bright spots or granules are created.

These granules and sunspots are features of the photosphere, which is a thin layer only 100 km thick. We are most familiar with this layer because it is the visible surface of the Sun. It produces most of the white light we see.

THE CHROMOSPHERE Above the surface photosphere is the chromosphere. The temperature rises from about 6,000°C to about 20,000°C. At this temperature, hydrogen emits a reddish light. Solar flares and eruptions are common in this region.

TRANSITION REGION Between the chromosphere and the corona is a thin, irregular layer that is poorly understood. This layer, called the transition region, is being examined by TRACE (Transition Region And Coronal Explorer). Within this region the temperature rapidly increases from 20,000°C to 1,000,000°C. Scientists are studying this region to increase their understanding of the processes that cause this temperature increase.

THE CORONA AND SOLAR WIND Above the chromosphere is the corona. The temperature in the corona is about 1,000,000°C. Hydrogen and other elements

Image of the Sun showing the interior, surface, corona, and solar wind.

are ionized and blown into space as a continuous outpouring of plasma known as the solar wind.

COMPOSITE IMAGE OF THE SUN

The interior image from MDI (on page 81) illustrates the rivers of plasma discovered flowing under the Sun's surface. The surface image was taken with the Extreme ultraviolet Imaging Telescope (EIT) at 304Å. Both were superimposed on a Large Angle Spectroscopic Coronograph (LASCO) C2 image, which blocks the Sun so that it can view the corona.

The image suggests the range of SOHO's (Solar and Heliospheric Observatory) research from the solar interior, to the surface and corona, and out to the solar wind.

TŁÉHONAA'ÉÍ / THE MOON

NAVAJO NAME: Tłéhonaa'éí
PRONUNCIATION: Tlay hon aa ay
TRANSLATION: Moon

WESTERN NAME: Moon
LOCATION: Moon
OBSERVED PATH: East to West

TŁÉHONAA'ÉÍ

Tłéhonaa'éí is considered to be a female energy in relation to the male energy of the Sun. There are many Navajo stories about the Moon, often passed down through the female side of a family.

The Moon is usually depicted as a perfect white shell disc, carried by the Moon carrier riding a horse across the sky. The phases of the Moon are considered to be very important, and there is a descriptive Navajo name for each of the 29 nights of the lunar cycle. The cycles of the Moon provide a timekeeping device for the days of the month.

Navajos begin counting the first day of the lunar cycle with dahiitą́, the first crescent Moon visible after the new Moon. Whereas the Western term for the new Moon refers to the night where no Moon is visible, the equivalent Navajo term, "new monthly cycle," dahiitą́, refers to the first crescent Moon visible after the moonless night.

Typically twelve full Moons can be seen each year, but every few years, the Navajo acknowledge a thirteenth Moon that must be added to the annual cycle to reset the calendar system so that the Sun and Moon cycles remain harmoniously balanced.

The Moon is closely watched for indication of weather to come. It is said that if the crescent Moon is tilted, with the points of the crescent toward the top of the sky, it is holding its contents and there will be little precipitation in the month to come. On the other hand, if the crescent Moon is standing upright, or the points are facing sideways or downward, it is losing its contents and rain or snow will be coming.

Full Moon.

The crescent Moon will first be seen setting in the Western sky after sunset, and approximately two weeks later, the full Moon, haníbą́ą́z, will first be seen rising in the east.

Each month, the full Moon can be seen in the sky at the same time as the Sun. In the evening, the Moon sits on the eastern horizon at the same time that the Sun can be seen on the western horizon. This event is called ahinéél'íí, and it is said that the Sun and the Moon are looking at each other.

The meaning of the Moon is life itself. It is intrinsically connected with the life cycles of organisms. The Moon is a regulator of many things: weather, ocean tides, calendar months, planting, life and birth, cycles of women, ceremonial functions, and behavioral influences.

Ceremonies are often planned to coincide with the phases of the Moon. When the Moon is growing it is called waxing. When the Moon is shrinking it is called waning. There are certain ceremonies that are done only when the Moon is waxing. This has to do with regeneration and growth.

The names of the months in Navajo are indicative of the natural cycles of animal, plant, and human life experiences associated with each month. Because

Galileo false-color composite image of the Moon.

the Navajo months are traditionally coordinated to the Moon's cycles, the time when a Navajo month begins may not be the same as when the Western months begin. See the full list of the Navajo Moon/month names on page 85.

THE MOON: SELENE

The Greeks called the Moon Selene, meaning light. Today, we use a variation of the name Selene in the name of the silvery white mineral selenium. When selenium is polished it shines and shimmers, much like moonlight.

The word "month" comes from a much older word, *Moonths,* referring to the cyclical pattern and phases of the Moon. The ancient Romans called the Moon Luna. This is where the words "lunar" and "lunatic" come from.

The Moon has many aspects in many cultures. It can represent birth, death, sacrifice, rebirth, and immortality. It also relates to the cycles of women, the tides, rain, and growth of vegetation. All of these contain aspects of transformation and cyclical change, related to the various phases of the Moon as observed from Earth every night.

To the ancient Greeks, the Moon was a female, a manifestation of Artemis, the goddess of the hunt and of the forest. There are many stories about the exploits of Artemis, in her guise as a skillful huntress. Other Greek stories refer to the Moon as having the ideal qualities of a woman.

Some Greek stories tell the origins of the mythical werewolf. The cult of Lycos in ancient Greece revered the wolf. The leader of this cult, Lycam, was the father of fifty wolf sons. The cult also revered the divine She-Wolf, a kind of lunar goddess. The She-Wolf was related to the werewolf, who was celebrated on nights with a full Moon.

Another Greek story about the Moon tells about the three Moiras. The Greek word Moira refers to a phase. These were the three phases of the Moon. The three Moiras were the New Moon, also known as the Virgin Goddess of Spring, the Full Moon, or Goddess of Summer, and the Waning Moon, or Old Goddess of Autumn.

CALENDRICAL COMPARISON

Navajo Months	Western Months	Western Full Moon Names (from the Farmer's Almanac)
Yasniłt'eez (Boiling Ice)	January	Wolf Moon
Atsá Biyaazh (Little Eagles)	February	Snow Moon
Wóózhch'íįd (Cry of Little Eagles)	March	Worm Moon
Tą́ą́chil (Little Leaves)	April	Pink Moon
T'ą́ą́soh (Large Leaves)	May	Flower Moon
Ya'iishjáátshchili (Small Planting)	June	Strawberry Moon
Ya'iishjáátsoh (Large Planting)	July	Buck Moon
Bini'ant'ą́ą́ts'ózi (Little Harvest)	August	Sturgeon Moon
Bini'ant'ą́ą́tsoh (Large Harvest)	September	Harvest Moon
Ghą́ą́jí (Parting of Seasons)	October	Hunter's Moon
Niłch'its'osí (Little Winds)	November	Beaver Moon
Niłch'itsoh (Big Winds)	December	Cold Moon

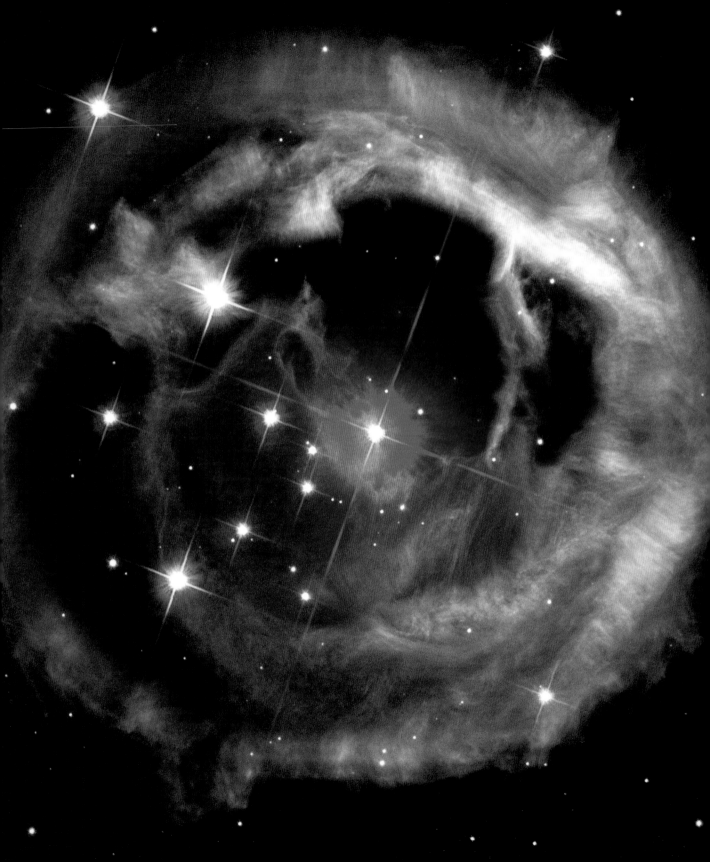

PART III:

RESOURCES

✦
✦
✦

GLOSSARY

Asterism: A small group of stars.

Black Hole: A place in space only a few miles in diameter from which no matter or radiation, including light, can escape because of the intense gravity of a dense collapsed star at its center.

Buckskin: The skin of a deer or elk, soft and pliable, often used for clothes or pouches.

Cardinal directions: East, south, west, and north are the four directional principal points. Sometimes referred to as the Four Directions.

Celestial: Relating to the sky, often as something heavenly or divine.

Centaur: A mythical man in ancient Greek culture, whose body is half man and half beast.

Ceremonial: System of formal rites and rituals according to tribal procedure and protocol.

Circumpolar: Referring to stars that revolve around the North Star and, for the most part, remain above the horizon.

Comparison: Placing together or juxtaposing two or more things to establish their similarities or dissimilarities.

Complementarity: The quality or state of being complementary. Often used in Navajo philosophy to describe the interrelationship of two entities which are dependent upon, supplement, or complete the perfection of a whole.

Constellation: A formation of stars perceived as a figure or design, especially one of the 88 recognized groups named after characters from classical Greek mythology and other cultures of the Mediterranean, as well as various common animals and objects.

Cosmology: Philosophy that deals with the character of the Universe as an orderly and harmonious system, particularly related to the processes of nature and interrelation of natural order.

Cosmography: A science that describes and maps the main features of the sky and the earth, including astronomy, geography, and geology.

Culture: A pattern of human behavior and characteristics including beliefs, social forms, and material traits constituting a distinct complex of tradition of a tribal or social group.

Diné: The name by which Navajo people refer to themselves, meaning "The People."

Diyin Diné'é: Spiritual or supernatural essences or beings, sometimes referred to as holy people (spirits) by the Navajo.

Dwarf star: A star of low luminosity and relatively small mass and size.

Elliptical: An elongated circle or oval. In ancient Greek, it meant "defective," or not quite a circle.

Equinox: Either of the two points at which the Sun crosses the celestial equator. The vernal equinox, or spring equinox, occurs when the Sun crosses the equator from south to north on or near

March 21 each year. The autumnal equinox, or fall equinox, occurs on or near September 21 when the Sun crosses the equator from north to south.

Galaxy: One of billions of large systems of stars including not only stars but nebulae, star clusters, globular clusters, and interstellar matter that make up the Universe.

Heliacal: Relating to or near the Sun

Heliacal Rise: The first rising of a star in the east during the predawn hour at which time it becomes visible to the observer (see page 71).

Heliacal Set: The last setting of a star in the west around sunset at which time it becomes invisible to the observer.

Hibernate: To pass the winter in a resting state, where the body temperature drops to a little above freezing and metabolic activity is reduced to nearly zero.

Hogan: A traditional Navajo home, usually octagon-shaped.

Holistic: Emphasizing the organic relation between parts and wholes, where determining factors in nature are wholes which are irreducible to the sum of their parts.

Hologram: A negative produced by exposing a high-resolution photographic plate, without camera or lens, near a subject illuminated by monochromatic, coherent radiation as from a laser. When it is placed in a beam of coherent light, a true three-dimensional image of the subject is formed.

Horizon: Based on the Greek word *horizein,* meaning "to separate" or "to define." Used to demarcate the boundary between earth and sky.

Hubble Space Telescope: The largest astronomical telescope ever put into space, which has taken photographs far out into the Universe, greatly expanding our knowledge.

Igniter: One that ignites, used in this case to describe a Navajo star that gives light and life to a nearby constellation.

Ionize: An atom or molecule that has lost or gained one or more electrons, thus becoming electrically charged.

Latitude: Angular distance north or south from the Earth's equator, used to locate and map specific areas.

Light year: Distance light travels in one year at a rate of 186,282 miles per second, about six trillion miles.

Longitude: Angular distance as measured from the zero meridian, used in geography and astronomy mapping and measurements.

Lunar: Related to the Moon.

Mediterranean: The lands and peoples around the Mediterranean Sea in southern Europe.

Metaphysical: Related to that which is conceived of as transcendant; deals with that which is beyond the physical or experimental.

Millennium: A period of 1,000 years. (Plural: **Millenia**)

Nebula: Bodies of rarified gas or dust in the interstellar space within a galaxy, including our own Milky Way, that often have a cloudy appearance.

Observation: The act of watching, noting, or regarding with attention for scientific or other special purpose.

Orbit: A path outlined by one body or object as it revolves around another.

Parallel: Forming lines in the same direction but not meeting.

Planisphere: A small-scale star map of the whole northern or southern sky with an oval or circular aperture that can be adjusted for any time and date, to show constellations which are, at that particular time and date, in the sky above the horizon.

Pollution: In astronomy, the term **light pollution** is used to describe man-made light that limits visibility in the night sky.

Protocol: An established code prescribing correct procedure.

Radiation: The process of emitting radiant energy in the form of waves or particles; the energy of electomagnetic waves.

Revolution: Orbital motion around a point.

Revolve: To go around or cause to go around in an orbit.

Rotate: To turn or cause to turn around an axis or a center.

Rotation: The act or process of turning around an axis or a center.

Solstice: From the Latin, meaning "the Sun stands still." One of two points on the ecliptic at which the Sun's distance from the celestial equator is greatest and which is reached by the Sun each year around June 21 (summer solstice) and December 21 (winter solstice).

Spiritual: Sacred.

Sumerians: Habitants of Sumer, the ancient region of lower Babylonia, in present-day Iraq.

Supernova: A massive star that has exploded, releasing energy making it millions of times brighter than the Sun.

Velocity: Speed, distance traveled per period of time.

Wholeness: Comprising the whole, entire or total. An unbroken completeness.

NAVAJO PRONUNCIATION GUIDE

The column on the left shows the approximate Navajo diacritical pronunciation marks. The column on the right offers an approximate phonetic pronunciation for the reader who cannot interpret the Navajo diacritical marks. Some Navajo sounds do not have formal equivalents in the English language.

1. High tone on all vowels that are marked with an accent: á, é, í, and ó.
2. Nasalization to all vowels that are marked with a hook: ą, ę, į, and ǫ.
3. Glottal stop, a closing of the throat, for all vowels with an apostrophe: '

NAVAJO	PHONETIC
Náhookǫs Bi'ką'	Na hoe kos Bih ka
Náhookǫs Bi'áád	Na hoe kos Bih aad
Náhookǫs Bikǫ'	Na hoe kos Bih kwo (*nasalize kwo*)
Dilyéhé	Dil yeh heh
Átsé Ets'ózí	A tseh Ets ozi
Hastiin Sik'aí'ií	Hasteen Sick eye ee
Gah Hahat'ee	Gaa ha haa tay
Yikáísdáhá	Yih kise daahaa
Ma'ii Bizǫ'	Mah ee Bihzon
Ii'ni	Ee neeh
Shash	Shash
Tsetah Dibé	Tseh tah dibeh
Tłish Tsoh	Tli sh tso
Ma'ii	Mah ee
Johannaa'éí	Jo haan aa ay
Tł'éhonaa'éí	Tlay hon aa ay

CONSTELLATION LOCATION GUIDE

This guide shows the *approximate locations* of the main Navajo constellations using the corresponding Greek constellations.

NAVAJO CONSTELLATIONS	WESTERN CONSTELLATIONS
Náhookǫs Bi'ką' (Revolving Male, sometimes referred to as Male Revolving One)	Big Dipper
Náhookǫs Bi'ááá (Revolving Female, sometimes referred to as Female Revolving One)	Cassiopeia
Náhookǫs Bikǫ' (Central Fire)	Polaris
Dilyéhé (Seed-like Sparkles)	Pleiades
Átsé Ets'ózí (First Slender One)	Orion
Hastiin Sik'aí'íí (Man with a Solid Stance, sometimes referred to as Man with Legs Ajar)	Corvus
Átsé Etsoh (First Big One)	Top half of Scorpius
Gah Hahat'ee (Rabbit Tracks)	Lower curved hook of Scorpius
Yikáísdáhá (Awaits the Dawn)	Milky Way
Ma'ii Bizǫ' (Coyote Star)	Canopus, seen in the south
Ii'ni (Thunder)	Pegasus makes up the body, while six other stars, beginning with Denebola, in Leo, outline the feather
Shash (Bear)	Sagittarius
Tsetah Dibé (Mountain Sheep)	Beehive Cluster in Cancer
Tłish Tsoh (Big Snake)	Canis Major and Puppis
Johannaa'éí (Sun)	The Sun
Tł'éhonaa'éí (Moon)	The Moon

SELECTED BIBLIOGRAPHY

Allen, Richard Hinckley. *Star Names: Their Lore and Meaning*. New York, NY: Dover Publications, 1963.

Aveni, Anthony. *Stairways to the Stars: Skywatching in Three Great Ancient Cultures*. New York, NY: John Wiley and Sons, 1997.

Bazin, Maurice, Modesto Tamex, and the Exploratorium Teacher Institute. *Math and Science Across Cultures: Activities and Investigations from the Exploratorium.* New York, NY: The New Press, 2002.

Berns, Barbara Brauner, and Associates. *Guiding Curriculum Decisions for Middle-Grades in Science.* Portsmouth, NH: Education Development Center, 2001.

Bruchac, Joseph. *Pushing Up the Sky: Seven Native American Plays for Children.* New York, NY: Dial Books for Young Readers, 2000.

Bruchac, Joseph, and Jonathan London. *Thirteen Moons on Turtle's Back: A Native American Year of Moons.* New York, NY: Philomel Books, 1992. Bruchac, Joseph, and Gayle Ross. *The Story of the Milky Way: A Cherokee Tale.* New York, NY: Dial Books for Young Readers, 1995.

Bryan, E.H. *Stars Over Hawaii* (2002 revisions and additions by Richard A. Crowe, PhD). Hilo, Hawaii: Petroglyph Press, 2002.

Caduto, Michael, and Joseph Bruchac. *Keepers of the Night: Native American Stories and Nocturnal Activities for Children.* Golden, CO: Fulcrum Publishing, 1994.

Farrar, Claire R. *Living Life's Circle: Mescalero Apache Cosmovision.* Albuquerque, NM: University of New Mexico Press, 1991.

Flood, Bo, William Flood, and Beret E. Strong. *Pacific Island Legends: Tales from Micronesia, Melanesia, Polynesia and Australia.* Honolulu, Hawaii: The Bess Press, 1999.

Galat, Joan Marie. *Dot to Dot in the Sky: Stories in the Stars.* Vancouver, BC: Whitecap Books, 2003.

Gardner, Robert. *Science Project Ideas About the Sun.* Berkeley Heights, NJ: Enslow Publishers, 1997.

Garza, Dolly. *Tlingit Moon and Tide: Teaching Resource, Elementary Level.* Fairbanks, AK: Alaska Sea Grant College Program, 1999.

Giles, Paulette. *North Spirit: Travels Among the Cree and Ojibway Nations and Their Star Maps.* Scarborough, ON: Doubleday Canada, Ltd., 1996.

Goldsmith, Donald. *Connecting with the Cosmos: Nine Ways to Experience the Wonder of the Universe.* Naperville, IL: Sourcebooks, Inc., 2002.

Griffen-Pierce, Trudy. *Earth Is My Mother, Sky Is My Father: Space, Time, and Astronomy in Navajo Sandpainting.* Albuquerque, NM: University of New Mexico Press, 1992.

Herlihy, Anna Friedman, and Elizabeth A. Kessler. *Astronomy Inspirations: A Guide to Art at the Adler.* Chicago, IL: Adler Planetarium and Astronomy Museum.

Keams, Geri. *Grandmother Spider Brings the Sun: A Cherokee Story.* Flagstaff, AZ: Northland Publishing Company, 1995.

Krupp, E. C. *Beyond the Blue Horizon: Myths and Legends of the Sun, Moon, Stars, and Planets.* New York, NY: Oxford University Press USA, 1991.

MacDonald, John. *The Arctic Sky: Inuit Astronomy, Star Lore, and Legend.* Iqaluit, NU: Nunavut Research Institute, 2000.

Malville, J. McKim, and Claudia Putnam. *Prehistoric Astronomy in the Southwest.* Boulder, CO: Johnson Books, 1989.

Maryboy, Nancy C. *A Guide to Navajo Astronomy.* Somerville, MA: Learning Technologies, Inc., 2004.

McCue, Harvey, and Associates, for the Department of Indian Affairs and Northern Development. *The Learning Circle: Classroom Activities on First Nations in Canada, ages 4–7.* Ottowa, ON: Published under the authority of the Minister of Indian Affairs and Northern Development.

Miller, Dorcas S. *Stars of the First People: Native American Star Myths and Constellations.* Boston, MA: Houghton Mifflin Company, 1997.

Milord, Susan. *Tales of the Shimmering Sky: Ten Global Folktales with Activities.* Charlotte, VT: Williamson Publishing, 1996.

Monroe, Jean Guard, and Ray A. Williamson. *They Dance in the Sky: Native American Star Myths.* Boston, MA: Houghton Mifflin Company, 1987.

Moroney, Lynn. *Moontellers: Myths of the Moon from Around the World.* Flagstaff, AZ: Northland Publishing Company, 1995.

Raymo, Chet. *365 Starry Nights: An Introduction to Astronomy for Every Night of the Year.* Upper Saddle River, NJ: Prentice Hall, 1990.

Sesti, Giuseppe Maria. *The Glorious Constellations: History and Mythology.* (From the Italian *Le Dimore Del* Cielo) New York, NY: Harry N. Abrams, Inc., 1990.

Shenandoah, Joanne, and Douglas M. George. *Skywoman: Legends of the Iroquois.* Santa Fe, NM: Clear Light Publishers, 1998.

Simon, Seymour. *Look to the Night Sky.* New York, NY: Puffin Books, 1979.

Snowden, Sheila. *The Young Astronomer.* Eveleth, MN: Usborne Books, 1983.

Souza, D. M. *Northern Lights* (ages 9–14). Minneapolis, MN: Carolrhoda Books, 1993.

Starr, Eileen M. *The World of the Maya.* Somerville, MA: Learning Technologies, Inc., 2003.

Thompson, Vivian. *Hawaiian Myths of Earth, Sea, and Sky.* Honolulu, HI: University of Hawaii Press, 1996.

Velarde, Pablita. *Old Father Story Teller.* Santa Fe, NM: Clear Light Publishers, 1989.

Walker, Christopher, ed. *Astronomy Before the Telescope.* London: British Museum Press, 1996.

Williamson, Ray A., and Claire R. Farrar. *Earth and Sky: Visions of the Cosmos in Native American Folklore.* Albuquerque, NM: University of New Mexico Press, 1992.

ACKNOWLEDGMENTS

GRATITUDE AND APPRECIATION TO:

Paul Soderman, Cathie Quigley-Soderman, and Chris O'Brien, World Hope Foundation; NASA Public Outreach and Educational Resources; NASA Goddard Sun-Earth Connection Forum; NASA Johnson Space Flight Center, office of Astronaut Appearances; Isabel Hawkins, University of California, Berkeley, Center for Science Education, Space Sciences Laboratory; Philip J. Sakimoto, Department of Physics, University of Notre Dame; Marv Bolt, Adler Planetarium & Astronomy Museum, Chicago; Northern Arizona University, Department of Physcis and Astronomy; NASA Arizona Space Grant; Space Telescope Science Institute; Utah Humanities Council; Charles Redd Center for Western Studies; Learning Technologies, Inc. (Starlab); Twin Rocks Trading Post, Bluff, Utah; Ashley Teren, www.sharingtheskies.com; Christopher Teren, Teren Solutions; and our deep gratitude to the respected Navajo cultural specialists who wish to remain anonymous, and whose input greatly enriched this work.

ART AND PHOTOGRAPHY CREDITS

front cover, page 24: Spiral Galaxy, NASA, Gerald Cecil (University of North Carolina), Sylvain Veilleux (University of Maryland), Joss Bland-Hawthorn (Anglo-Australian Observatory), and Alex Filippenko (University of California at Berkeley), Image Type: Astronomical STScI-PRC2001-28

front cover left inset: Frank Zullo, Comet Hale-Bopp over petroglyphys, Picacho Mountains, Arizona.

front cover middle inset: Navajo Constellation painting by Melvin Bainbridge (Navajo Nation)

front cover right inset, page 29: Hubble (HST) Courtesy NASA/Space Telescope Science Institute

back cover, page 32: Hubble Heritage (STScI) Team Robert O'Dell (Rice University), Thomas P. Ray (Dublin Institute for Advanced Study), and David Corcoran (University of Limerick)

pages 2–3, 39: Star Trails Above Mauna Kea (Rob Ratowsky) Gemini Observatory / Travis Rector, University of Alaska, Anchorage

pages 4–5, 63: HD 10146, NASA, ESA, D. R. Ardila (JHU), D. A. Golimowski (JHU), J. E. Krist (STScI/JPL), M. Clampin, (NASA/GSFS), J. P. Blakeslee (JHU), H. C. Ford ((JHU), G. F. Hartig (STScI), G. D. Illingworth (UCO-Lick) and the ACS Science Teampage 6: Hubble Site/NASA/ESA, NGC 3370

page 9: Diné Navajo Worldview conceptualized by David Begay and Nancy C. Maryboy, design by Ken Grett, Northern Arizona University, Flagstaff, AZ

page 10: Diné Universe conceptualized by David Begay and Nancy C. Marboy, painted by Melvin Bainbridge, printed with permission of the Indigenous Education Institute.

page 13: Ant Nebula STScI-PRC2001-05

page 12: Map by Karen Schober, Seattle, WA

page 14: Navajo Hogan with Enhanced Skies: Hogan photo by Nancy C. Maryboy, image design by Troy D. Cline, Goddard Space Flight Center, NASA, Galaxy image credit to NGC 4214, Image type: Astronomical, Image credit NASA and Hubble Heritage, Team STScI

page 15: Cosmic Planning Model Image, conceptualized by David Begay and Nancy C. Maryboy with design by Christopher S. Teren

page 16: Navajo Basket, photo by Nancy C. Maryboy

page 17: Reprinted with permission from Map of the Universe (The Northern Hemisphere) by Howard Figler, 2000, 1979, Celestial Arts, Berkeley CA

page 18: Contemporary basket by Lorraine Black, courtesy of Twin Rocks Trading Post, Bluff, Utah

page 19: Greek Temple image is provided royalty-free for educational noncommercial settings only, copyright 2000–2005, All Rights Reserved by Eric Rymer

page 22: Armillary sphere image, courtesy of Adler Planetarium and Astronomy Museum, Chicago, Illinois

page 27: Kim Dismukes, John Ira Petty

page 30: Hubble Heritage team, Image Type: Astronomical STScI-PRC2005-15, Y. H. Chu and R. M. Williams (UIUC)

page 33: Artist's conception taken by permission from Ed Krupp's diagram (*Beyond the Blue Horizon: Myths and Legends of the Sun, Moon, Stars, and Planets*, 1991)

pages 34, 36, 38, 41, 45, 49, 51, 53, 55, 57, 64 top, 66 top, 70 top, 72, 77, and 82: Navajo Constellation paintings by Melvin Bainbridge (Navajo Nation)

pages 35, 37, 46, 50, 52, 60, 61, 64 bottom, 66 bottom, 70 bottom, and 73: Greek images by Johann Hevelius, Atlas, 1690

page 42: Pleiades, NASA/ESA

page 44: NASA and The Hubble Heritage team (STScI/AURA) George Herbig and Theodore Simon (Institute for Astronomy, University of Hawaii)

page 48: NASA, NOAO, ESA and the Hubble Heritage team (STScI/AURA) K. Noll, C. Luginguhl, F. Hamilton, Image Type: Astronomical STScI-PRC2001-12

page 54: ESA/NASA and Felix Mirabel (French Atomic Energy Commission and Institute for Astronomy and Space Physics / Argentina)

56: Barney Magrath, courtesy NASA/GSFC APOD

page 62: NASA Galaxy Evolution Explorer (GALEX) satellite

page 65: NASA, ESA, S. Beckwith (STScI) and the Hubble Heritage Team (STScI/AURA) Image Type: Astronomical STScI-PRC2005-12a

page 67: NASA, ESA and J. Hester (Arizona State University)

page 68 right: Hubble and Spitzer Space Telescopes Artist's View

page 68 left: NASA, ESA, STScI, J. Hester and P. Scowen (Arizona State University)

page 74: Photo credit: Jon Morse (University of Colorado), and NASA Investigating Team: Kris Davidson (University of Minnesota), Bruce Balick (University of Washington), Dennis Ebbets (Ball Aerospace), Adam Frank (University of Minnesota), Fred Hamann (University of California, San Diego), Roberta Humphreys (University of Minnesota), Sveneric Johansson (Lund Observatory), Jon Morse (University of Colorado), Nolan Walborn (Space Telescope Science Institute), Gerd Weigelt (Max Planck Institute for Radio Astronomy, Bonn), and Richard White (Space Telescope Science Institute) Image Type: Astronomical STScI-PRC1996-23a

page 76: NASA, ESA, STScI, J. Hester and P. Scowen (Arizona State University) Image Type: Astronomical STScI-PRC1995-44a

The following images are Courtesy of SOHO consortium. SOHO is a project of international cooperation between ESA and NASA

    page 79 top: EIT 04A

    page 79 bottom: He II at 304A, SOHO/EIT, SOHO/LASCO

    page 80: diagram SOHO (ESA & NASA)

    page 81: LASCO C2 image, SOHO (ESA & NASA)

pages 83, 84: Dave Williams and Jay Friedlander, NASA Goddard Space Flight Center

page 86: NASA image

page 94: Hubble Space Telescope image of Galaxies, Artwork by Theresa Breznau, Living Earth Studios, Bluff Utah, 2007

All Navajo Constellation Paintings © copyright of Indigenous Education Institute.

Together, we are all looking at the same cosmic essence, enriched by cultural diversity and many ways of knowing. Hubble Space Telescope image of galaxies. Artwork by Theresa Breznau, Living Earth Studios, Bluff, Utah, 2007.

Rio Nuevo Publishers®
P. O. Box 5250
Tucson, AZ 85703-0250
(520) 623-9558, www.rionuevo.com

Fourth Edition © 2010 by Nancy C. Maryboy, PhD, and David H. Begay, PhD
(First Edition © 2005, Second Edition © 2007, Third Edition © 2008)

See page 93 for photography and illustration credits.

Library of Congress Cataloging-in-Publication Data

Maryboy, Nancy C.
Sharing the skies : Navajo astronomy / Nancy Cottrell Maryboy and David Begay ; original concept co-designed by Indigenous Education Institute, Bluff, Utah, and World Hope Foundation, Boulder, Colorado. — 4th ed.
    p. cm.
Includes bibliographical references.
ISBN-13: 978-1-933855-40-0 (pbk. : alk. paper)
ISBN-10: 1-933855-40-1 (pbk. : alk. paper) 1. Indian astronomy—Southwest, New. 2. Navajo Indians—Science. 3. Navajo Indians—Folklore. 4. Stars—Mythology—Southwest, New. 5. Astronomy—Cross-cultural studies. I. Begay, David. II. Indigenous Education Institute. III. World Hope Foundation. IV. Title.
E98.A88M37 2009
979.1004'9726—dc22

                                    2009028316

Printed in China
Book design: Karen Schober, Seattle, Washington

10 9 8 7 6 5